Secrets of the Aurora Borealis

By Dr. Syun-Ichi Akasofu

In collaboration with Jack Finch and Jan Curtis

ALASKA GEOGRAPHIC® / Vol. 29, No. 1 / 2002

To teach many more to better know and more wisely use our natural resources...

EDITOR
Penny Rennick

PRODUCTION DIRECTOR
Kathy Doogan

ASSOCIATE EDITOR
Susan Beeman

MARKETING DIRECTOR
Mark Weber

ISBN: 1-56661-058-3

PRICE TO NON-MEMBERS THIS ISSUE: $23.95

PRINTED IN U.S.A.

POSTMASTER
Send address changes to:

ALASKA GEOGRAPHIC®
P.O. Box 93370, Anchorage, AK 99509-3370

BOARD OF DIRECTORS
Kathy Doogan
Carol Gilbertson
Penny Rennick

Robert A. Henning, **PRESIDENT EMERITUS**

COVER: *An auroral curtain, typically about one mile thick, curves over moonlit mountains in the Alaska Range. See page 103 for details on how this photo was taken. (Patrick J. Endres)*

PREVIOUS PAGE: *Like a fire in the sky, the red aurora struck terror into the hearts of early people. Scientists now know that red auroras tend to occur during years with high sunspot activity. (Greg Syverson)*

FACING PAGE: *Auroras can be visible during various times throughout the night, from early evening to dawn. (Daryl Pederson)*

ALASKA GEOGRAPHIC® (ISSN 0361-1353) is published quarterly by The Alaska Geographic Society, 639 West International Airport Rd. #38, Anchorage, AK 99518. Periodicals postage paid at Anchorage, Alaska, and additional mailing offices. Copyright © 2002 The Alaska Geographic Society. All rights reserved. Registered trademark: Alaska Geographic, ISSN 0361-1353; key title Alaska Geographic. This issue published March 2002.

THE ALASKA GEOGRAPHIC SOCIETY is a non-profit, educational organization dedicated to improving geographic understanding of Alaska and the North, putting geography back in the classroom, and exploring new methods of teaching and learning.

MEMBERS RECEIVE ALASKA GEOGRAPHIC®, a high-quality, colorful quarterly that devotes each issue to monographic, in-depth coverage of a specific northern region or resource-oriented subject. Back issues are also available (see page 112). Membership is $49 ($59 to non-U.S. addresses) per year. To order or to request a free catalog of back issues, contact: Alaska Geographic Society, P.O. Box 93370, Anchorage, AK 99509-3370; phone (907) 562-0164 or toll free (888) 255-6697, fax (907) 562-0479, e-mail: akgeo@akgeo.com. A complete list of our back issues, maps and other products can also be found on our website at www.akgeo.com.

SUBMITTING PHOTOGRAPHS: Those interested in submitting photos for possible publication should write or refer to our website for a list of upcoming topics or other photo needs and a copy of our editorial guidelines. We cannot be responsible for unsolicited submissions. Please note that submissions must be accompanied by sufficient postage for return by priority mail plus delivery confirmation.

CHANGE OF ADDRESS: When you move, the post office may not automatically forward your *ALASKA GEOGRAPHIC®* issues. To ensure continuous service, please notify us at least six weeks before moving. Send your new address and membership number or a mailing label from a recent issue of *ALASKA GEOGRAPHIC®* to: Address Change, Alaska Geographic Society, P.O. Box 93370, Anchorage, AK 99509-3370.

If your issue is returned to us by the post office because it is undeliverable, we will contact you to ask if you wish to receive a replacement for a small fee (to cover the cost of additional postage to reship the issue).

PRE-PRESS: Graphic Chromatics
PRINTING: Banta Publications Group / Hart Press

The Library of Congress has cataloged this serial publication as follows:

Alaska Geographic. v.1-
[Anchorage, Alaska Geographic Society] 1972-
v. ill. (part col.). 23 x 31 cm.
Quarterly
Official publication of The Alaska Geographic Society.
Key title: Alaska geographic, ISSN 0361-1353.

1. Alaska—Description and travel—1959-
—Periodicals. I. Alaska Geographic Society.
F901.A266 917.98'04'505 72-92087
Library of Congress 75[79112] MARC-S.

ABOUT THIS ISSUE

Dr. Syun-Ichi Akasofu, founding director of the International Arctic Research Center (IARC) in Fairbanks, is one of a handful of distinguished experts on the aurora. In this issue of *ALASKA GEOGRAPHIC®* Dr. Akasofu brings to life myths and facts about the northern lights based on years of intense research.

Dr. Akasofu received his Bachelor of Science in 1953 and his Master of Science in 1957 from Tohoku University in Japan, then came to the Geophysical Institute (GI) at the University of Alaska Fairbanks (UAF) the next year as a graduate student. By 1961 he'd earned a Ph.D., and in 1986 he became director of the GI. He taught geophysics at UAF and in 1999 was appointed director of the IARC. His auroral research has formed the basis of the discipline of magnetospheric substorms.

He has also won numerous awards.

Photographers Jack Finch and Jan Curtis have worked closely with Dr. Akasofu to produce aurora images and gather data. Jack Finch lives in Fairbanks. Many of his aurora photos supplement the GI website. Jan Curtis is a former staff member of the GI. His aurora images have appeared in numerous books, magazines, and articles. He is currently the state climatologist for Wyoming.

Dr. Akasofu would like to thank his colleagues — senior, contemporary, and junior — who have advanced the understanding of the aurora. Without their efforts, this book wouldn't have been possible. Thanks also go to Kimberly Hayes for administrative support and Tohru Saito for help with diagrams.

EDITOR'S NOTE: Terms shown in color and boldface appear in the glossary on page 106.

Contents

 # Introduction

But, where, O Nature, is thy law?
From the midnight lands comes up the dawn!
Is it not the sun setting his throne?
Is it not the icy seas that are flashing fire?
Lo, a cold flame has covered us!
Lo, in the night-time day has come upon the
* Earth.*

What makes a clear ray tremble in the night?
What strikes a slender flame into the
* firmament?*
Like lightening without storm clouds,
Climbs to the heights from Earth?
How can it be that frozen steam
Should midst winter bring forth fire?

(Mikhail Vasil'evich Lomonosov,
1743, in Sydney Chapman's
"History of Aurora and Airglow," 1966)

The aurora borealis — the northern lights — is one of the most spectacular natural phenomena on Earth. Named for Aurora, the rosy-fingered goddess of the dawn in Roman mythology who heralded the rising sun, its beauty and splendor are often beyond description. Charles F. Hall,

an author and polar explorer of the nineteenth century, simply exclaimed with a sigh, "Who but God can conceive such infinite scenes of glory? Who but God could execute them, painting the heavens in such gorgeous display?" William H. Hooper, another polar explorer, reported, "Language is vain in the attempt to describe its ever varying and gorgeous phases; no pen nor pencil can portray its fickle hues, its radiance, and its grandeur." Even so, many of those characters passing through the history of the Arctic and Antarctic regions, among them adventurers, scholars, miners, settlers, and gamblers, made an attempt to communicate in words to their less fortunate fellows the wonder they witnessed.

Although no description of the aurora can do it full justice, its magnificence is at least suggested in the above poem by the eighteenth century Russian scientist Mikhail Vasil'evich Lomonosov.

This book is an attempt to present the whole story of the aurora. The aurora borealis has intrigued mankind since ancient times. Early descriptions of the

aurora are contained in the Old Testament, in the mythology of the Lapps, the Eskimos and the American Indians, as well as in Medieval European literature. After sampling some of these accounts, we shall proceed to the narratives of polar explorers and tales of settlers in the northern lands.

The aurora has fascinated many famous philosophers and scientists, among them, Aristotle, Descartes, Goethe, Henry Cavendish, John Dalton, Edmund Halley, Alfred L. Wegener, and Benjamin Franklin. Since the days of these men, the aurora has provided one of the most challenging problems encountered in modern science. In these pages, you will be introduced to the history of auroral science and the development of our understanding of the phenomenon, and we'll explain, in layman's language, what we now know about how the aurora works and also what we do not. ■

FACING PAGE: *Viewed from directly beneath an auroral curtain, the display looks like emanating rays. (Greg Syverson)*

# Auroral Legends

Eskimo Legends

Virtually every northern culture has its oral legends about the aurora, passed down for generations. The Eskimos, Athabaskan Indians, Lapps, Greenlanders, and even the Northwest Indian tribes were familiar with this mysterious light in the sky. Their legends took many forms and were most often associated with their notions of life after death. Typical Eskimo stories on the aurora may be found in books by Knud Rasmussen and Ernest W. Hawkes, explorers and anthropologists who set to paper oral traditions. In his book, *The Labrador Eskimo* (1916), Hawkes related one legend:

The ends of the land and sea are bounded by an immense abyss, over which a narrow and dangerous pathway leads to the heavenly regions. The sky is a great dome of hard material arched over the Earth. There is a hole in it through which the spirits pass to the true heavens. Only the spirits of those who have died a voluntary or violent death, and the raven have been over this pathway. The spirits who live there light torches to guide the feet of new arrivals. This is the light of the aurora. They can be seen there feasting and playing football with a walrus skull.

The whistling crackling noise which sometimes accompanies the aurora is the voices of these spirits trying to communicate with the people of the Earth. They should always be answered in a whispering voice. Youths and small boys dance to the aurora. The heavenly spirits are called selamiut, sky dwellers, 'those who live in the sky.'

Rasmussen's story in *Intellectual Culture of the Iglulik Eskimo* (1932) follows the same general theme, even though it is taken from another Eskimo culture:

The aurora borealis is called arsharneq or arshät. It is personified in a powerful spirit who is in great demand as a helping spirit for the best shamans. We believe that the aurora borealis is alive just as men and women are; for if you whistle at it, it crackles and comes nearer. But if you spit at it, it all runs together in the middle and forms another picture. It is just as if it understood people and did what they wanted it to do....

... The dead suffer no hardship, wherever they may go, but most prefer nevertheless to dwell in the Land of Day, where the pleasures appear to be without limit. Here, they are constantly playing ball, the Eskimos' favourite game, laughing and singing, and the ball they play with is the skull of a walrus. The object is to kick the skull in such a manner that it always falls with the tusk downwards, and thus sticks fast in the ground. It is this ballgame of the departed souls that appears as the aurora borealis, and is heard as a whistling, rustling, crackling

FACING PAGE: *A geomagnetic storm on March 31, 2001, created this purple aurora above the Talkeetna Mountains in Broad Pass, just east of Denali National Park and Preserve. (Fred Hirschmann)*

sound. The noise is made by the souls as they run across the frost-hardened snow of the heavens. If one happens to be out alone at night when the aurora borealis is visible, and hears this whistling sound, one has only to whistle in return and the light will come nearer, out of curiosity.

Indian Legends

American Indians, also, have legends about the aurora. From Katherine B. Judson's *Myths and Legends of the Pacific Northwest* (1910):

Chief M'Sartto, Morning Star, had an only son, different from all others in the tribe. The son would not play with the other children but would take his bow and arrow and be away for days. Curious as to what mischief his boy could be up to, the Chief, one day, followed him. His journey progressed and all at once a queer feeling filled the old chief, as if all knowledge was floating away from him. His eyes suddenly closed and when they were opened he found himself in an extremely light country with no sun, moon, or stars. There were many people

about, and they spoke a strange language that he did not understand. The people were engaged in a wonderful game of ball that seemed to turn the light to many colors. The players all had lights on their head and wore very curious belts called Menquan, or Rainbow belts. After many days of searching for his son, the old Chief met a man who spoke his language. The man had also traveled by chance to this strange new country and knew of the Chief's son. When he was brought to his son, the Chief saw him playing ball with the others, and strangely enough the boy's light was brighter than any there. When the game was ended, the old Chief was introduced to the people and honored by the Chief of the Northern Lights; two great birds were ordered to be brought forth, K'che Sippe by name. On these birds the two dwellers of the Lower World returned home from the Wa-ba-ban, the land of the Northern Lights, following the Spirits Path, Ket-a-gus-wowt or Milky Way. Again all knowledge was erased from Chief M'Sartto. When they arrived home, the Chief's wife paid no notice that they were gone because she was afraid that they would never return and was very relieved. So it is that those very few who travel to the Land of the Northern Lights do not remember their

The auroral curtain, the basic form, often appears in multiples, such as these three green bands seen from Denali Park. The Big Dipper is also visible in the upper left corner of this image. (John W. Warden)

Auroral rays appear to descend through clouds. Legends, many concerned with spirits of the dead, abound about mysteries of the aurora. (Daryl Pederson)

remarkable journey. But the Chief's son remembers and travels there often.

In *The Mythology of All Races* (1964), Louis Herbert Gray described an Indian story of the aurora:

From the abode of the Cannibal, the Kwakiutl say, red smoke arises. Sometimes the "cannibal pole" is the rainbow, rather than the Milky Way; but the Cannibal himself is regarded as living at the north end of the world (as is the case with the Titanic beings of many Pacific-Coast myths), and it is quite possible that the Cannibal is originally a war-god typified by the Aurora Borealis. A Tlingit belief holds that the souls of all who meet a violent death dwell in the heaven-world of the north, ruled by Tahit, who determines those that shall fall in battle, of what sex children shall be born, and whether the mother shall die in child-birth. The Aurora is blood-red when these fighting souls prepare for battle, and the Milky Way is a huge tree-trunk (pole) over which they spring back and forth. [Anthropologist Franz] Boas is of opinion that the secret societies originated as warrior fraternities among the Kwakiutl, whose two most famed tutelaries are the Cannibal and Winalagilis, the Warrior of the North. Ecstasy is supposed to follow the slaying of foe; the killing of a slave by the Cannibal Society members is in a sense a celebration of victory, since the slave is war booty; and it is significant that in certain tribes the Cannibals merely hold in their teeth the heads of enemies taken in war.

Scandinavian Legends

In the thirteenth-century Norwegian Chronicle, *The King's Mirror* (a Viking story), we find the king instructing his son about the aurora:

No man is sure what those lights can be which the Greenlanders called the Northern Lights, even those who have spent a long time in Greenland. Of course, thoughtful men will make conjectures as to what they might consist of. Their nature is peculiar in that the darker the night is, the brighter they seem and they always appear at night, never by day. In appearance they resemble a vast flame of fire viewed from a great distance. It also looks as if sharp points were shot from this flame up into the sky of uneven height and in constant motion, now one,

now another darting highest; and the light appears to blaze like a living flame. When these rays are at their highest, people out of doors can easily find their way about and can even go hunting. Sometimes, however, the light appears to grow dim, as if a black smoke or a dark fog were blown up among the rays; and it looks very much as if the light were overcome by this smoke and about to be quenched. It happens at times that people think they see large sparks shooting out of the light from a glowing iron that has just been taken from the forge. As night declines, the light begins to fade and when daylight appears, it seems to vanish entirely.

Men who have thought about the origin of these lights have guessed at three sources, one of which ought to be the true one. Some hold that fire circles about the ocean and all the bodies of water that stream about on the outer sides of the globe; since Greenland lies on the outermost edge of the earth to the north, they think it possible that these lights shine forth from the fires that encircle the outer ocean. Others say that during the hours of night, when the sun's course is beneath the earth, an occasional gleam of its light may shoot up into the sky; for they insist that Greenland lies so far out on the earth's edge that the curved surface which shuts out the sunlight must be less prominent there. Still others believe that the frost and the glaciers have become so powerful there that they are able to radiate forth these flames.

Gray described a mythological tale of Finnish Lapps:

According to the Finnish Lapps the Aurora Borealis is 'the dead in battle, who, as spirits, still continue battling with one another in the air.' The Russian Lapps also declare the Aurora Borealis to be 'the spirits of the murdered.' These live in a house, in which at times they gather together and begin stabbing one another to death, covering the floor with blood.

'They are afraid of the sun, hiding themselves from its rays.' The Aurora Borealis appears 'when the souls of the

According to a Finnish legend, the aurora is created when the fox named Repu splashes snow into the air with his long tail. (By Sheré Chamness, courtesy of Robert Eather; from S.-I. Akasofu)

murdered begin their slaughter.' Hence the Lapps fear it.

The Estonians also see in the Northern Lights a heavenly war, 'Virmalised taplevad' ('Virmalised fight'). On the island of Ösel they say that during the holy nights when the heavens open, one may see two armed fighting-men, eager to give battle to one another, but God will not allow it, and separates them. Most probably the Finns also possessed a similar belief; in certain Karelian magic songs Pohjola is sometimes mentioned as the residence of those who 'were killed without sickness' and where the inhabitants are said to have 'blood-dripping garments.' In a variation he wakes from his trance, and the 'fish' has brought him back uninjured to his body again. Jessen is able further to affirm that 'the louder a shaman can sing, the longer is his snake.' The Finns also relate in their tales how the Lapps fly in the shape of birds through the air; when one of these is shot down, the Lapp tumbles to the ground. These soul-animals were sometimes pictured on the magic drums.

These legends illustrate the basic human desire to make sense of natural phenomena like the aurora. ◼

One of the earliest records of auroral sightings was described in a Babylonian tablet reporting red glow in the northern sky (about 350 B.C.). One can also find descriptions of the aurora as far back as the days of the Roman Empire, in Seneca's *Questiones Naturales*, and later in Aristotle's *Meteorologica*. Aristotle described the aurora as glowing air which was released from cracks, called "chasms," of the sky. It was quite natural to consider the aurora as such, because stars were considered to be holes in the heavens. He considered that comets were a similar phenomenon. Seneca, a Roman philosopher at the beginning of the Christian era, noted:

There are chasmata (fissures), when a certain portion of the sky opens, and gaping displays the flame as in a torch. The colours also of all these are many. Certain are of the brightest red, some of a flitting and light flame-colour, some of a white light, others shining, some steadily and yellow without eruptions or rays.

There are a number of biblical references to phenomena that can be attributed to auroral activity. In the Old Testament one example is in 2 Maccabees, chapter 5, verses 1-4, written about 176 B.C.:

A medieval artist's view of the aurora borealis over the Bavarian city of Bamberg in 1560 shows that active auroras, appearing often as swift streams of light, were associated with battles fought in the sky. (Courtesy of Dept. of Prints and Drawings, Zentral Bibliothek, Zürich; from S.-I. Akasofu)

About this time Antiochus sent his second expedition into Egypt. It then happened that all over the city, for nearly forty days, there appeared horsemen charging in midair, clad in garments interwoven with gold — companies fully armed with lances and drawn swords; squadrons of cavalry in battle array, charges and countercharges on this side and that, with brandished shields and bristling spears, flights of arrows and flashes of gold ornaments, together with armor of every sort. Therefore all prayed that this vision might be a good omen.

There were a number of auroral sightings reported in both China and Japan. One Chinese example (translated from *Sung-shih*) is:

Red cloud spreading all over the sky, and among the red cloud there were ten-odd bands of white vapour like glossed silk penetrating it. They arose from Tzu-wei, invading the Great Dipper and the Wen-chang, and then dispersed from the southeast.

"Swords," "spears," "white vapor, like glossed silk penetrating it," are all likely

Red auroras were considered a sign of ill omens during medieval times and pilgrimages were organized to avert the wrath of Heaven. Brilliant displays have frightened people as recently as this century in regions where aurora sightings are rare. (Courtesy of Zentral Bibliothek, Zürich; from S.-I. Akasofu)

what we call the ray structure in the red aurora. Such descriptions make their account more credible in terms of auroral sighting than a description like "just a red glow."

On many occasions in historical times the aurora has reached as far south as the middle latitudes, and it has struck fear into the populations of Italy and France. Such an aurora is rich in dark red color, and the people of Europe associated it with blood and battle. It was an ill omen of disasters or a war to come. Accounts of such red auroras can be found in some of the earliest of Eastern and Western writings. Seneca remarked:

Amongst these we may notice, what we frequently read of it in history, the sky is seen to burn, the glow of which is occasionally so high that it may be seen amongst the stars themselves, sometimes so near the Earth that it assumed the form of a distant fire. Under Tiberius

Caesar the cohorts ran together in aid of the colony of Ostia as if it were in flames, when the glowing of the sky lasted through a great part of the night, shining dimly like a vast and smoking fire.

In his monograph, *The Aurora Borealis* (1897), Alfred Angot, honorary meteorologist at the central meteorological office in France, described the terror caused by red auroras in medieval times:

… astrology has so troubled the minds of men that the aurora borealis had become a source of terror: bloody lances, heads separated from the trunk, armies in conflict, were clearly distinguished. At the sight of them people fainted (according to Cornelius Gemma), others went mad. Pilgrimages were organized to avert the wrath of Heaven, manifested by these terrible signs. Thus, according to the Journal of Henri III, in the month of September 1583, eight or nine hundred persons of all ages and both sexes, with their lords, came to Paris in procession, dressed like penitents or pilgrims, from the villages of Deux-Gemeaux and Ussy-en-Brie, near La Ferte-Gaucher, 'to say their prayers and make their offerings in the great church at Paris; and they said that they were moved to this penitential journey because of signs seen in heaven and fires in the air, even towards the quarter of the Ardenned, whence had come the first such penitents, to the number of ten or twelve thousand, to Our Lady Pheims and to Liesse.' The chronicler adds that this pilgrimage was followed a few days afterwards by five others, and for the same cause.

It is somewhat surprising to find a painting from medieval days that depicts the aurora as a series of candles in the sky, but it is quite likely that such a romantic view of the aurora was not very common. ◼

The Aurora and Polar Explorers

After the great voyages of exploration by such men as Vasco da Gama and Ferdinand Magellan, the polar regions became the remaining frontier for explorers, adventurers, and travelers. Many of them were astounded by the beauty of the aurora. Their encounters with the phenomenon were described prolifically in their memoirs, narratives, and ships logs. Among these explorers were Frederick Cook, Sir William E. Parry, Sir John Franklin, Charles Hall, Adolphus W. Greely, Elisha Kent Kane, Fridtjof Nansen, Adolf E. Nordenskiöld, Capt. James Cook, Robert F. Scott, and Roald Amundsen.

Frederick Cook, who claimed to have reached the North Pole before Robert E. Peary, described his encounter with the aurora in *My Attainment of the Pole* (1911):

We continued our course, the Eskimos singing, the dogs occasionally barking. Hours passed. Then we all suddenly became silent. The last, the supreme, glory of the North flamed over earth and frozen sea. The divine fingers of the aurora, that unseen and intangible thing of flame, who comes from her mysterious throne to smile upon a benighted world, began to touch the sky with glittering, quivering lines of glowing silver. With skeins of running, liquid fire she wove over the sky a shimmering panorama of blazing beauty. Forms of fire, indistinct and unhuman, took shape and vanished. From horizon to zenith, cascades of milk-colored fire ascended and fell, as must the magical fountains of heaven.

In the glory of this other-world light I felt the insignificance of self, a human unit; and, withal I became more intensely conscious than ever of the transfiguring influence of the sublime ideal to which I had set myself. I exulted in the thrill of an indomitable determination, that determination of human beings to essay great things — that human purpose which, throughout history, has resulted in the great deeds, the great art, of the world, and which lifts men above themselves. Spiritually intoxicated, I rode onward. The aurora faded. But its glow remained in my soul.

Tragic Exploration of the Northwest Passage

Out of the countless polar expeditions, the quest for the Northwest Passage, a shortcut from Europe to the Orient across the Canadian polar wilderness, perhaps led most to the kindling of curiosity about the aurora. The passage itself was believed to exist by those of the opinion that if it was possible for one to pass around the southern tip of South America, one should likewise be able to sail around the northern part of North America. The search was begun by Sir Martin Frobisher who explored the southern part of Baffin Island (Frobisher Bay) in 1576-1578. He thought that one side of the bay was the Asian

FACING PAGE: *A major solar flare in November 2001 created spectacular auroral displays that kept photographers out all night, witnesses to the dancing light rays and stars. Orion's belt seems to float just above the mountains in Hatcher Pass, 40 miles north of Anchorage. (Greg Syverson)*

continent and the other side the American continent. He returned with a large amount of pyrite (fool's gold) after misidentifying it as gold. He was followed by many famous explorers, whose names remain on islands, straits, and bays in this area, among them Davis, Hudson, Baffin, and Parry.

Capt. William Parry was one of the first to reach the Arctic Ocean in his search for the Northwest Passage (but not the Bering Sea). Anyone who has witnessed auroral displays would agree with the explorer that it is impossible to record them adequately in words:

Thus far description may give some faint idea of this brilliant and extraordinary phenomenon, because its figure here

maintained some degree of regularity; but during the most splendid part of its continuance, it is, I believe, almost impossible to convey to the minds of others an adequate conception of the truth. It is with much difference, therefore, that I offer [a] description, the only recommendation of which perhaps is, that it was written immediately after witnessing this magnificent display. ...

(From *Journal of a Second Voyage for the Discovery of a North-West Passage from the Atlantic to the Pacific; Performed in the Years 1821-22-23, in His Majesty's Ships* Fury *and* Hecla, 1904)

During the middle of the nineteenth century, it was found that Lancaster Sound was the entrance into the Northwest Passage, but the British Admiralty became

FAR LEFT: *The disappearance of experienced British Arctic explorer Sir John Franklin and his crew aboard the ships* Terror *and* Erebus *in 1846 led to several descriptions of auroral sightings by search parties in northern Canada. (Courtesy of Alaska and Polar Regions Dept., Elmer E. Rasmuson Library, University of Alaska Fairbanks; from S.-I. Akasofu)*

BELOW: *Every man aboard Sir John Franklin's doomed ships trapped in ice off Cape Felix in the Northwest Territories perished while awaiting rescue. Franklin had described the aurora he saw during an earlier expedition as "beams, flashes, and arches," while those who searched for him called the mysterious lights the "glowing heavens," and a "wondrous phenomenon." (Courtesy of Alaska and Polar Regions Dept., Elmer E. Rasmuson Library, University of Alaska Fairbanks; from S.-I. Akasofu)*

Red auroras, seen here as the corona form, occur higher in the atmosphere than the more common green displays, making them visible farther from the poles. In 1958, people as far south as Mexico reported seeing a red aurora borealis. (Greg Syverson)

impatient at the continued failure to discover the Northwest Passage. The British decided to send their most experienced Arctic explorer, Sir John Franklin, with a crew of 129 officers and men in Her Majesty's ships *Terror* and *Erebus* to solve this problem once and for all. They left London in 1845 and were seen last by a British whaling ship when they were passing through Lancaster Sound. Franklin's journey turned out to be one of the most tragic of all polar expeditions. The ships were beset by ice in M'Clintock Channel and all members of the party eventually perished after months of hardship and starvation. An old Eskimo woman who had seen them later in their journey reported that they "fell down and died as they walked along."

In the ensuing years the absence of any trace of Franklin's expedition resulted in the mounting of an intensive search, including a series of expeditions financed by Lady Franklin. Finally, in 1859, a short message was found in a small cairn by Capt. Francis L. McClintock and his search party at King William Island that reads (from Leslie H. Neatby's *In Quest of the Northwest Passage*, 1958):

April 25, 1848. — H.M. ships *Terror* and *Erebus* were deserted on the 22nd of April, 5 leagues NNW, of this, having been beset since 12th September, 1846. The officers and crews, consisting of 105 souls, under the command of Captain F.R.M. Crozier, landed here in lat. 69 37 42 N., long. 98 41 W. Sir John Franklin died on the 11th June, 1847; and the total loss by death in the expedition has been to this date 9 officers and 15 men.

Numerous memoirs were written by search party leaders describing their fruitless search activity. Men on these expeditions, wintering in the Canadian wilderness, witnessed the aurora and were astonished by its beauty. It is partly through the extraordinary circumstance of the great search that the brilliance of the aurora became widely known to the civilized world at that time.

Sir John Franklin himself was a good observer of the aurora. Although his own memoirs of this ill-fated expedition were never recovered, he left two voluminous reports of his earlier overland expedition from the western shore of Hudson Bay to the Arctic Ocean. In that account he gave the following description:

For the sake of perspicuity I shall describe the several parts of the Aurora,

Some auroras display only muted colors, such as this one over Fairbanks. Auroral lights are emitted by atoms and molecules in the upper atmosphere when a beam of high-speed electrons hits them. Energy from solar flares, and in turn, solar wind, is directly connected to auroral activity. (Gary Schultz)

it revolves itself into beams which, by a quick undulating motion, project themselves into wreaths, afterwards fading away, and again and again brightening without any visible expansion or contraction of matter. Numerous flashes attend in different parts of the sky.

(From the 1823 book *Narrative of a Journey to the Shores of the Polar Sea in the Years 1819, 1820, 1821, 1822*)

One of the many Franklin search party leaders, Charles Hall was a religious person. He visited many Native villages, hoping to find a few survivors in them. During the winter, he was amazed by the beauty of the aurora and in his 1864 book *Arctic Researches and Life Among the Esquimaux: Narrative of an Expedition in Search of Sir John Franklin, in the Years 1860, 1861, and 1862*, he exclaimed:

Again, on another morning, December 17th, at six o'clock, I write: The heavens are beaming with aurora. The appearance of this phenomenon is quite changed from what it has been. Now the aurora shoots up in beams scattered over the whole canopy, all tending to meet at zenith. How multitudinous are the scenes presented in one hour of the aurora! This morning the changes are very rapid and magnificent. Casting the eye in one direction, I view the instantaneous flash of the aurora shooting up and spreading out its beautiful rays, gliding this way, then returning, swinging to and fro like the pendulum of a mighty clock. I cast my eyes to another point; there instantane-

which I term beams, flashes and arches.

The beams are little conical pencils of light, ranged in parallel lines, with their pointed extremities towards the Earth, generally in the direction of the dipping-needle.

The flashes seem to be scattered beams approaching nearer to the Earth, because they are similarly shaped and infinitely larger. I have called them flashes, because their appearance is sudden and seldom continues long. When the aurora first becomes visible it is formed like a rainbow, the light of which is faint, and the motion of the beams indistinguishable. It is then in the horizon. As it approaches the zenith

ous changes are going on. I close my eyes for a moment; the scene has changed for another of seemingly greater beauty. In truth, if one were to catch the glowing heavens at each instant now passing, his varied views would number thousands in one hour. Who but God could conceive such infinite scenes of glory? Who but God execute them, painting the heavens in such gorgeous display?

Three search party leaders later pointed to the inadequacy of words to describe the magnificent displays of the aurora. Lt. William H. Hooper wrote in *Ten Months Among the Tents of the Tuski, with Incidents of an Arctic Boat Expedition in Search of Sir John Franklin* (1853):

Few nights passed without a greater or less display of the Aurora Borealis, that wondrous phenomenon whose existence after more than half a century of research, is yet unaccounted for satisfactorily. Language is vain in the attempt to describe its ever-varying and gorgeous phases; no pen nor pencil can pourtray its fickle hues, its radiance, and its grandeur.

And from Adolphus W. Greely's *Three Years of Arctic Service, an Account of the Lady Franklin Bay Expedition of 1881-84 and the Attainment of the Farthest North* (1894):

The aurora of January 21st was wonderful beyond description, and I have no words in which to convey any adequate idea of the beauty and splendor of the scene. It was a continuous change from arch to streamers, from streamers to patches and ribbons, and back again to arches, which covered the entire heavens for part of the time. It lasted for about twenty-two hours, during which at no moment was the phenomena other than vivid and remarkable. At one time there were three perfect arches, which spanned the southwestern sky from horizon to horizon. The most striking and exact simile, perhaps, would be to liken it to a conflagration of surrounding forests as seen at night from a cleared or open space in their centre.

Elisha Kent Kane wrote in *U.S. Grinnell Expedition in Search of Sir John Franklin* (1854):

Soon from the circumference of this arch proceeded a fimbriated or fringy series of purple cirri, delicately tinted at their edges, increasing with wonderful regularity, and extending in long, ray-like processes of cloud to an altitude of some twenty degrees above the horizon. Before eleven o'clock, these processes had

Current-carrying electrons that excite atoms in the high atmosphere where the aurora occurs flow in a thin sheet or sheets, creating an aurora that looks like a curtain of lights in the sky. (John W. Warden)

become long, stratiform illuminated clouds, beautifully marked, of a breadth, measured roughly by the eye, of four or five degrees, interrupted where they crossed the illuminated region of the sun, but everywhere else extending over the heavens to the south and west (true); and although still diminishing in intensity, extending nearly to the eastern quarter of the sky. By coalescing at their bases, these radiating processes augmented the size of the central segment. The intervals between them appeared, by contrast, to be artificially illuminated.

Although the Northwest Passage was eventually penetrated from east to west by Amundsen in 1903, it is still one of the most difficult — and impractical — ocean passages in the world. The largest oil tanker with ice-breaking capabilities constructed in the United States, the *S.S. Manhattan*, traversed the passage in 1969 in a test of the feasibility of using that route to transport Alaska oil from Prudhoe Bay. The big ship, encountering great difficulty, barely made it through and the idea was abandoned.

One of the most important scientific contributions during the search for the Northwest Passage was the discovery of the magnetic pole by both John Ross and James Ross in 1831 during their magnetic survey at the Boothia Peninsula in Canada. The magnetic pole is the location where a magnetic compass needle stands perpendicularly when it is suspended and allowed to rotate freely in a vertical plane.

All compass needles point toward the magnetic pole. Scientists thought that there would be a magnetic mountain the size of Mt. Blanc there. However, they discovered that the point was simply a desolate seashore in the Arctic wilderness. The magnetic field of Earth is generated deep in its core, not by Earth's crust.

The magnetic field of Earth can be represented by a hypothetical bar magnet located at Earth's center. In an ideal situation, the bar magnet would be aligned along the rotation axis. Then, the geographic North Pole would be the south magnetic pole and the geographic South Pole would be the north magnetic pole. Actually, the axis of the hypothetical

From a pull-off along the Parks Highway in Southcentral Alaska, the aurora is much more accessible to the average person than is the summit of Mount McKinley, seen here in the distance at 20,320 feet. (Greg Syverson)

Earth's magnetic field, a factor in where the aurora occurs (at the poles), can be represented by a hypothetical bar magnet at the center of the planet. Magnetic field research has helped scientists better understand the aurora. (S.-I. Akasofu)

Fridtjof Nansen was, among other accomplishments, a talented artist and many of his books were illustrated by his own woodcuts and paintings. One woodcut depicts a majestic aurora and his ship *Fram*, which was designed by him to withstand the tremendous pressure of the Arctic pack ice. During his attempt to reach the North Pole, Justin F. Denzel, an associate, wrote of Nansen:

Shortly after midnight, Nansen left the party for a quiet, solitary stroll across the ice. It was a beautiful clear night, with the gaudy streamers of the aurora borealis shifting across the heavens. During his walk, he turned and looked back to see the dark masts and rigging of the Fram silhouetted against the pale yellow glow of the sky. Behind it, the silken draperies of light were shimmering across the heavens like great pulsating rays of violet sheen, intermingled with pastel shades of pink and green. The very air crackled with the brilliant iridescence, lending an eerie glow to the surrounding landscape.

For almost an hour, he stood there, spellbound by this gorgeous display. In the cold silence, he thought of the long months they had already spent in this vast frozen wasteland and the even longer months that might lay ahead. He thought, too, of Norway and of home, and for a fleeting moment he felt an agonizing pang of regret as he conjured up little Liv and Eva waiting for him on the shore. How long would it be before

bar magnet is offset from the rotation axis by about 11.5°. Thus, when the line containing the bar magnet is extended to Earth's surface, that particular point is located in Ellesmere Island, not at the geographic North Pole. This point is called the **geomagnetic pole**. The fact that the geographic pole and the geomagnetic pole do not coincide is due to distortions of Earth's magnetic field. In this book, we deal only with the geomagnetic pole. The magnetic field of Earth is one of the crucial elements that constitutes the generator that powers auroral electrical discharge.

he saw them again? With a shrug of his shoulders, he pushed the thought from his mind and turned his steps back to the Fram, to his merry companions and the warm comfortable cabins.

(From *Adventure North, The Story of Fridtjof Nansen*, 1968)

BELOW: *A triple-curtained aurora similar to Nansen's woodcut streaks the night sky. (By Takeshi and Aiko Matsuo; from S.-I. Akasofu)*

RIGHT: *In a woodcut, Fridtjof Nansen depicts himself strolling on the ice with a triple-curtained aurora overhead. (In Nansen's* Nord I Takeheimen, *1911, courtesy of University of Oslo; from S.-I. Akasofu)*

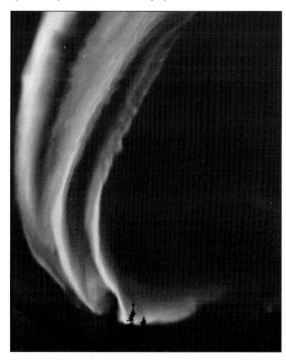

In the early twentieth century, explorer Roald Amundsen described Antarctica's aurora australis, or southern lights, as "Nature … in her best attire." (Courtesy of Alaska and Polar Regions Dept., Elmer E. Rasmuson Library, University of Alaska Fairbanks; from S.-I. Akasofu)

After Nansen's attempt to reach the North Pole, the sturdy *Fram* took Roald Amundsen to Antarctica in 1911, where he became the first man to reach the South Pole. Thus, this small ship holds the record for reaching the farthest north and farthest south of any single vessel.

Adolf E. Nordenskiöld

It appears that for polar explorers, the aurora was something unique to their life. Thus it is interesting to note the following written by geologist-explorer, Adolf E. Nordenskiöld, who made the first successful exploration of the Northeast Passage, the Arctic route from Gothenburg, Sweden, to Yokohama, Japan, along the Siberia coast in 1878-1879:

This splendid natural phenomenon … plays, though unjustifiably, a great role in imaginative sketches of winter life in the high north, and it is in the popular idea so connected with the ice and snow of the Polar lands, that most of the readers of sketches of Arctic travel would certainly consider it an indefensible omission if the author did not give an account of the aurora as seen from his winter station.

(From *The Voyage of the Vega Round Asia and Europe*, 1881)

Aurora and Antarctic Explorers

Famous explorer and navigator Capt. James Cook, who made three great voyages of discovery to the Pacific, is said to be the first European to witness the aurora in the Southern Hemisphere, the aurora australis. In his log of February 15, 1773, he wrote:

…we had fair weather, and a clear serene sky; and, between midnight and three o'clock in the morning, lights were seen in the heavens, similar to those in the northern hemisphere, known by the name of Aurora Borealis, or Northern Lights; but I never heard of the Aurora Australis being seen before. The officer of the watch observed, that it sometimes broke out in spiral rays, and in a circular form; then its light was very strong, and its appearance beautiful.

(From *A Voyage Towards the South Pole and Round the World*, 1777)

Capt. Cook saw the aurora borealis in the Pacific Ocean, on his way from the Sandwich Islands (Hawaii) to Alaska, but curiously, he made no mention of it when he was in Alaska waters.

Robert F. Scott, who died in an Antarctic blizzard in 1912 after his defeat by Amundsen in the race for the South Pole, described the aurora australis in his daily journal:

The eastern sky was massed with swaying auroral light, the most vivid and beautiful display that I had ever seen — fold on fold the arches and curtains of vibrating luminosity rose and spread across the sky, to slowly fade and yet again spring to glowing life.

The brighter light seemed to flow,

now to mass itself in wreathing folds in one quarter, from which lustrous streamers shot upward, and anon to run in waves through the system of some dimmer figure as if to infuse new life within it.

It is impossible to witness such a beautiful phenomenon without a sense of awe, and yet this sentiment is not inspired by its brilliancy but rather by its delicacy in light and colour, its transparency, and above all by its tremulous evanescence of form. There is no glittering splendor to dazzle the eye, as has been too often described; rather the appeal is to the imagination by the suggestion of something wholly spiritual, something instinct with a fluttering ethereal life, serenely confident yet restlessly mobile.

One wonders why history does not tell us of 'aurora' worshippers, so easily could the phenomena be considered the manifestation of 'god' or 'demon.' To the little silent group which stood at gaze before such enchantment it seemed profane to return to the mental and physical atmosphere of our house. Finally when I stepped within, I was glad that there had been a general movement bedwards, and in the next half-hour the last of the roysterers had succumbed to slumber.

(From Leonard Huxley's *Scott's Last Expedition, The Personal Journals of Captain R.F. Scott, R.N., C.V.O. on his Journey to the South Pole*, 1941)

In *The South Pole: An Account of the*

Norwegian Antarctic Expedition in the "Fram", 1910-1912 (1913), Roald Amundsen described the aurora australis by watching it from the *Fram*:

The light is so wonderful; what causes this strange glow? It is clear as daylight, and yet the shortest day of the year is at hand. There are no shadows, so it cannot be the moon. No; it is one of the few really intense appearances of the aurora australis that receives us now. It looks as though Nature wished to honour our guests, and to show herself in her best attire. And it is a gorgeous dress.

Modern-day aurora enthusiasts drop to their knees to photograph a northern lights display over Chena Hot Springs, a one-hour drive northeast of Fairbanks. Despite today's scientific explanation of the aurora, viewers still feel awe. (Gary Schultz)

Incidentally, at present a number of scientific research efforts are being coordinated at the South Pole, the U.S. Amundsen-Scott base. The aurora is continuously recorded by a number of instruments there, including an **all-sky camera.** ■

The Aurora and Northern Settlers

Northern Adventurers

Following the exploits of the polar explorers, it is natural that settlers, traders, and miners would flock to inhabit the new land. There were those who sought riches, particularly gold and furs, and those who wanted to start a new life with a healthy dose of adventure and the promise of a peaceful existence in the wilderness. These early settlers left numerous accounts of the aurora; it was a phenomenon little understood by them, but was frequently mentioned in their notes and letters. Verses were composed, perhaps inscribed by lantern or candlelight in the long darkness of the northern winter, when the cold allowed only a brief step out of the cabin's warmth to marvel at the wonder in the sky. Naturally, these witnesses viewed the aurora each in his own way, depending on their purpose for being in the Far North. Frederick Whymper, in *Travel and Adventure in the Territory of Alaska*, (1868) wrote:

Just as we were turning in for the night, a fine auroral display in the N.W. was announced, and we all rushed out to witness it from the roof of the tallest building in the Fort. It was not the conventional arch, but a graceful, undulating, ever-changing 'snake' of electric light; evanescent colours, pale as those of a lunar rainbow, ever and again flitting through it, and long streamers and scintillations moving upward to the bright stars, which distinctly shone through its hazy, ethereal form. The night was beautifully calm and clear, cold, but not intensely so, the thermometer at sixteen degrees [Fahrenheit].

As an example of a traveler's description, we may quote from several sources; following is an excerpt from one of the books by the modern Marco Polo — Bayard Taylor:

I … looked upward, and saw a narrow belt or scarf of silver fire stretching directly across the zenith, with its loose, frayed ends slowly swaying to and fro down the slopes of the sky. Presently it began to waver, bending back and forth, sometimes slowly, sometimes with a quick, springing motion, as if testing its elasticity. Now it took the shape of a bow, now undulated unto Hogarth's line of beauty [a serpentine line according to the aesthetics outlined in *The Analysis of Beauty* (1753) by William Hogarth], brightening and fading in its sinuous motion, and finally formed a shepherd's crook, the end of which suddenly began to separate and fall off, as if driven by a strong wind, until the whole belt shot away in long, drifting lines of fiery snow. It then gathered again into a dozen dancing fragments, which alternately advanced and retreated, shot hither

FACING PAGE: *This photo of the aurora shimmering over the Matanuska Valley and Pioneer Peak illustrates the difference in altitudes between the green and red auroras. The red color on top is called the* red line, *the green section the* green line. *The green color is from atomic oxygen ([O], separated oxygen molecules). (Cary Anderson)*

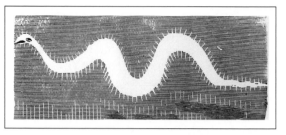

LEFT: *An aurora twists back and forth in this artist's version of the infamous northern lights over a fur trading post. (In Frederick Whymper's* Travel and Adventure in the Territory of Alaska, 1868, *courtesy of Alaska and Polar Regions Dept., Elmer E. Rasmuson Library, University of Alaska Fairbanks; from S.-I. Akasofu)*

ABOVE: *Some people once thought of the aurora as a radiating snake dancing in the sky. (In Elizabeth M'Hardie's* The Midnight Cry: "behold the bridegroom cometh," 1883, *courtesy of Alaska and Polar Regions Dept., Elmer E. Rasmuson Library, University of Alaska Fairbanks; from S.-I. Akasofu)*

and thither, against and across each other, blazed out in yellow and rosy gleams or paled again, playing a thousand fantastic pranks, as if guided by some wild whim.

We lay silent with upturned faces, watching this wonderful spectacle. Suddenly, the scattered lights ran together, as by a common impulse, joined their bright ends, twisted them through each other, and fell in a broad, luminous curtain straight downward through the air until its fringed hem swung apparently but a few yards over our heads. This phenomenon was so unexpected and startling, that for a moment I thought our faces would be touched by the skirts of the glorious drapery. It did not follow the spherical curve of the firmament, but hung plumb from the zenith, falling, apparently, millions of leagues through the air, its folds gathered together among the stars and its embroidery of flame sweeping the Earth and shedding a pale, unearthly radiance over the wastes of snow. A moment afterwards and it was again drawn up, parted, waved its flambeaux and shot its lances hither and thither, advancing and retreating as before. Anything so strange, so capricious, so wonderful, so gloriously beautiful, I scarcely hope to see again.

(From *Prose Writings of Bayard Taylor*, 1864)

And from George Kennan's *Tent Life in Siberia*... (1870):

Among the few pleasures which reward the traveler for the hardships and dangers of life in the far north, there are none which are brighter or longer remembered than the magnificent Auroral displays which occasionally illumine the darkness of the long polar night, and light up with a celestial glory the whole blue vault of heaven. No other natural phenomenon is so grand, so mysterious, so terrible in its unearthly splendor as this; the veil which conceals from mortal eyes the glory of the eternal throne seems drawn aside, and the awed beholder is lifted out of the atmosphere of his daily life into the immediate presence of God.

Edward S. Ellis wrote in *Among the Esquimaux; or Adventures Under the Arctic Circle* (1894):

At times these dartings resembled immense spears, and then they changed to bands of light, turning again into ribbons which shivered and hovered in the sky, with bewildering variation, turning and doubling upon themselves, spreading apart like an immense fan, and then trembling on the very verge of the horizon, as if about to vanish in the darkness of night.

During an expedition to Bossekop in Norway in 1837-1838, M. Bravais sketched this display, accurately illustrating main auroral features. (Courtesy of Alaska and Polar Regions Dept., Elmer E. Rasmuson Library, University of Alaska Fairbanks; from S.-I. Akasofu)

At the moment the spectators held their breath; fearing that the celestial display was ended; the streamers, spears, bands of violet, indigo, blue, orange, red, green, and yellow, with the innumerable shades, combinations, and mingling of colors, shot out and spread over the sky like the myriad rays of the setting Sun.

This continued for several minutes, marked by irregular degrees of intensity, so impressive in its splendor that neither lad spoke, for he could make no comment upon the exhibition, the like of which is seen nowhere else in nature.

But once both gave a sigh of amazed delight when a ribbon, combining several vivid colors, quivered, danced, and streamed far beyond the zenith, with a wary appearance that suggested that some giant, standing upon the extreme

northern point of the Earth, had suddenly unrolled this marvelous ribbon and was waving it in the eyes of an awestruck world.

One of the most striking features of that mysterious electrical phenomena known as the Northern Lights is the absolute silence which accompanies them. The genius of man can never approach in the smallest degree the beauties of the picture without some noise, but here nature performs her most wonderful feat in utter stillness. The panorama may unfold, roll together, spread apart again with dazzling brilliancy and suddenness, but the strained ear catches no sound, unless disassociated altogether from the phenomenon itself, such as the soft sighing of the Arctic wind over the wastes

of snow, or through the grove of solemn pines.

There were moments when the effluence spread over the Earth, like the rays of the midnight Sun, and the lads, standing in front of the primitive dwelling of the Esquimau, resembled a couple of figures stamped in ink in the radiant field.

For nearly an hour the rapt spectators stood near the entrance to the native dwelling, insensible to the extreme cold, and too profoundly impressed to speak or stir; but the heavens had given too great a wealth of splendor; brilliancy, color, and celestial scene-shifting to continue it long. The subtle exchange of electrical conditions must have reached something like an equipoise, and the overwhelming beauty and grandeur exhausted itself.

The ribbons and streamers that had been darting to and beyond the zenith, shortened their lightning excursions into space, leaping forth at longer intervals and to a decreasing distance, until they ceased altogether, displaying a few flickerings in the horizon, as though eager to bound forth again, but restrained by a superior hand with the command, 'Enough for this time.'

A pinkish aurora lights up the sky above a well-insulated cabin in southcentral Alaska. In 1862, explorer Bayard Taylor wrote of the aurora's "embroidery of flame sweeping the Earth and shedding a pale, unearthly radiance over the wastes of snow." (Greg Syverson)

Current understanding of the natural world reaches back to early polar explorers and settlers who took note of their surroundings in minute detail. This drawing presents auroral curtains. Contrary to popular belief in the early nineteenth century, which placed auroral activity highest near the North Pole, explorer Robert E. Peary noted that he'd seen auroras of "greater beauty" in Maine than he had north of the Arctic Circle. (In I.I. Hayes's Recent Polar Voyages, *1861, courtesy of Alaska and Polar Regions Dept., Elmer E. Rasmuson Library, University of Alaska Fairbanks; from S.-I. Akasofu)*

Fred [Warburton] drew a deep sigh.

'I never dreamed that anywhere in the world one could see such a sight as that.

'It is worth a voyage from home a hundred times over, and I don't regret our stay on the iceberg, for we would have been denied it otherwise.

'If there are any people living near the North Pole, it must be like dwelling in another world. I don't see how they stand it.

'I believe that the Northern Lights have their origin between here and the Pole,' said Fred; 'though I am not sure of that.

'The magnetic pole, which must be the source of the display, is south of the Earth's pole, and I suppose that's the reason for the belief you mention. But it is enough to fill one with awe, when he gazes on the scene and reflects that the world is one great reservoir of electricity, which, if left free for a moment by its Author, would shiver the globe into nothingness, and leave only an empty void where the Earth swung before.

'I pity the man who says 'There is no God,' or who can look unmoved to the very depths of his soul by such displays of infinite power.'

And finally, during his trip to Iceland, Charles S. Forbes witnessed the aurora and described it in *Iceland: Its Volcanoes, Geysers, and Glaciers* (1860):

The evening was very enjoyable after the hurricane, and twilight was relieved by a most brilliant aurora, which in these high latitudes often follows or precedes any great change of temperature; its pale, sylph-like, and undulating rays flitted about in every direction, and were only extinguished in one quarter of the heavens to be rekindled more brilliantly in another. Gradually increasing in power, its light equalled that of the moon, and, together with the intensity of the atmosphere, threw the distant western peaks apparently at our feet, and distorted them with inconceivable rapidity into fantastic and fairy forms, making inexhaustible demands on the

*The **red lower border**, different from the red line, is emitted by excited nitrogen molecules (N_2) at the bottom folded curtain of an active aurora. This photo was taken in the Alaska Range in March 1999. (Tom Walker)*

> But I'll tell you now — and if I lie, may
>   my lips be stricken dumb —
> It's a mine, a mine of the precious stuff
>   that men call radium.
> It's a million dollars a pound, they say,
>   and there's tons and tons in sight.
> You can see it gleam in a golden stream
>   in the solitudes of night.
> And it's mine, all mine — and say!
>   If you have a hundred plunks to spare,
> I'll let you have the chance of your life,
>   I'll sell you a quarter share.

In one of his most famous poems, "The Cremation of Sam McGee," he wrote:

> There are strange things done in the
>   midnight sun
> By the men who moil for gold;
> The Arctic trails have their secret tales
> That would make your blood run cold;
> The Northern Lights have seen queer
>   sights,
> But the queerest they ever did see
> Was that night on the marge of Lake
>   Labarge
> I cremated Sam McGee.

eye and imagination. Brilliant coronas from time to time encircled our zenith, but the climax was attained when a stream of light, rising in the west, seemed to unfold itself from the conical … and, graciously but slowly advancing, arched the heavens, bisected the pale Road of Winter....

The Aurora and Early Gold Miners

The gold rush in Canada and Alaska brought a great number of gold miners who dreamed of striking it rich. In spite of the effort required to survive in the harsh winters, many left written materials about the aurora.

Some gold miners thought the aurora was a vapor from a hidden mine. Robert Service presented this view tongue-in-cheek in his poem, "The Ballad of the Northern Lights":

> Some say that the Northern Lights are the
>   glare of the Arctic ice and snow;
> And some that it's electricity, and nobody
>   seems to know.

Another example of the outpouring of verse by the settlers and gold miners in Alaska and the Yukon is from Frank B. Camp's book, *Alaska Nuggets* (1922):

I'm a Sour-dough of Alaska
Who knows the Give and Take;
At Dawson, Nome, Iditarod,
I grabbed a handsome stake,
Then mushed it to the Outside
With all my Golden Pokes,
And invested them in business
That throttles me and chokes.

Today while working in my chair,
In office high up in the air,
Surcharged with smoke and other hazes;
My vision played a trick on me,
And set my eyes a roaming free,
Till once again I glimpsed the Wonder
    Blazes.

I saw the Blaze of Alaska,
    on the Blaze was a picture dear,
And the streaming haze of the housetop
    was parted till it was clear.
It showed me the Trails now hidden,
    it pictured the mountains high,
The river sands, where with their hands,
    men pan for gold till they die.
It revealed the swamps and the forests,
    where the big moose live and the bear;
Where the ptarmigans nest, on the high
    hill's crest, where the black fox has
    his lair.
It painted the bright aurora,
    that shines with a myriad lights,

As it brightens the snow, where the North
    winds blow, chill on the winter nights.

Jeremiah Lynch, in *Three Years in the Klondike* (1904), described well the atmosphere of a gold mining town:

Here was a smoking, drinking, gambling crowd of men mingled with gaudily-dressed girls dancing in the rear and drinking in the front. It was very close, and yet the keen icy frost outside was so penetrating that, even though the double doors were opened but for an instant for someone going or coming, the air of the place was not at all uncomfortable, despite the fierce and vivid heat from the giant stove. A dozen large lamps, each of which consumed a gallon of oil every twelve hours, gave a fairly good light, not much clearer, however, than the marvelous Aurora Borealis outside, lighting up the heavens from horizon to horizon with many-coloured hues. After midnight, those who had cabins went to them over the ice trail reluctantly and shiveringly, those who had not, slept on and under the billiard-tables, leaving narrow passages along the sides through which men could walk. These had no other home, nor could the town provide one, so perforce they lived

Klondike gold miner Jeremiah Lynch described auroral rays as "lighting up the heavens from horizon to horizon with many-coloured hues," as this auroral curtain does, reflected in Knik River north of Anchorage. (Cary Anderson)

The northern lights inspired literature, from lines in Robert Service's poems "The Ballad of the Northern Lights" and "The Cremation of Sam McGee," to a scene in Jack London's 1927 The Call of the Wild. *(Cary Anderson)*

thus, eating where and when they could and might. For the winter had brought destitution to many who had remained and the authorities found it necessary to provide work for those who needed sustenance.

In a famous gold mining story by Jack London, *The Call of the Wild* (1927), Buck, a dog that was brought to the Canadian wilderness from the south by his owner, eventually joined in with a pack of wolves. The book ends:

… But he is not always alone. When the long winter night comes on and the wolves follow their meat into the lower valleys, he may be seen running at the head of the pack through the pale moonlight or glimmering borealis…. ■

A large number of scientists and amateurs were attracted by the beauty of the aurora and speculated about its causes. Not knowing the characteristics of the auroral lights, however, their ideas remain only of historical interest. J.J.D. de Mairan, a French scientist, speculated that the aurora appeared when materials in the zodiacal light (a faint glow surrounding the inner solar system) were attracted by Earth's gravitational force and burned. He published the first scientific book on the aurora. Following are ideas proposed by Edmond Halley, Benjamin Franklin, Sophus Tromholt, Thomas W. Knox, and Karl Selim Lemström.

Edmond Halley (1656-1742)

Halley is well known as the discoverer of Halley's Comet and as the person who financed Isaac Newton's great treatise *Philosophiae Naturalis Principia Mathematica*. When he was 60, Halley was fascinated by an auroral display on the evening of March 6, 1716, and published a paper. He began with his detailed observation and noticed that a curtainlike form of the aurora tends to incline along the direction of a magnetic needle suspended so it is able to rotate freely in the vertical plane. He was familiar with Earth's magnetic field because the British Admiralty had assigned him the task of surveying it for their navigational purposes.

Halley speculated that the aurora was caused by water vapor that pours out from Earth during earthquakes and is ignited to radiate. However, he realized that locations of earthquakes are too limited to explain the aurora. Thus, he conceived what he called "magnetic atoms" that are abundant within Earth. He said that Earth's crust is porous to magnetic atoms, so that they can freely come out and radiate while moving along the direction of a magnetic line of force. When a bar magnet is placed below a sheet of paper on which fine iron filings

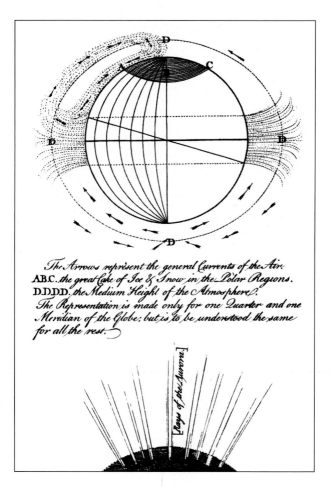

Benjamin Franklin's diagram shows his conception of the circulation of Earth's atmosphere and its relation to the aurora. (Courtesy of Franklin Institute; from S.-I. Akasofu)

are spread, the iron filings tend to line up along a number of curves, which connect both ends of the magnetic pole. These lines are called the magnetic line of force. A compass needle lines up along it. Halley's idea was later extended by J.B. Biot and others who conjectured that the magnetic atoms are ejected by volcanic eruptions.

Benjamin Franklin (1706-1790)

Since Benjamin Franklin is well known as an experimenter who examined electricity by flying a kite in a thunderstorm, it was natural for him to speculate about the cause of the aurora. In his "Suppositions and Conjectures Towards Forming an Hypothesis for its Explanation," he described his idea:

> If in the middle of a room you heat the Air by a Stove, a Pot of burning Coals near the Floor, the heated Air will rise to the Ceiling [sic], spread there over the cooler Air till it comes to the cold Walls; there being condensed and made heavier, it descends to supply the Place of that cool Air which had moved towards the Stove or Fire, in order to supply the Place of the heated Air which had ascended from the Space around the Stove or Fire.
>
> Thus there will be a continual Circulation of Air in the Room, which may be render'd visible by making a little Smoke; for that Smoke will rise and circulate with the Air.
>
> A similar Operation is perform'd by Nature on the Air of the Globe. Our Atmosphere is of a certain height,

This nineteenth-century engraving shows the corona, seen when the auroral curtain is located a little south of its zenith. (In Sophus Tromholt's Under the Rays of the Aurora Borealis, *1885, courtesy of Elmer E. Rasmuson Library, University of Alaska Fairbanks; from S.-I. Akasofu)*

This photo mirrors the form seen in polar explorer Tromholt's engraving. (By Jack Finch; from S.-I. Akasofu)

perhaps at a Medium Miles. Above that height it is so rare as to be almost a Vacuum. The Air heated between the Tropics is continually rising, its Place is supply'd by northerly and southerly Winds which come from the cooler regions.

The light, heated Air, floating above the cooler and denser, must spread northward and southward, and descend near the two poles, to supply the Place of the cooler Air which had moved towards the Equator.

Thus a circulation of Air is kept up in our Atmosphere as in the Room above mentioned.

That heavier and lighter Air may move in Currents of different and even opposite direction, appears sometimes by the Clouds that happen to be in those Currents, as plainly as by the Smoke in the Experiment above mentioned. Also, in opening a Door between two Chambers, one of which has been warmed, by holding a candle near the top, near the bottom, and near the middle, you will find a strong current of warm air passing out of the warmed Room above, and another of cool Air entering it below, while in the Middle there is little or no Motion.

The great quantity of Vapour rising between the Tropics forms Clouds, which contain much electricity. ...

If the Rain be received in an isolated Vessel, the Vessel will be electrified; for every drop brings down some Electricity with it.

The same is done by Snow and Hail. ...

If the Clouds are not sufficiently discharg'd by this Means, they sometimes discharge themselves by striking into the Earth, where the Earth is fit to receive their Electricity.

... Snow falling upon frozen Ground has been found to retain its Electricity; and to communicate it to an isolated Body, when after falling, it has been driven about by the Wind.

The Humidity, contain'd in all the equatorial Clouds that reach the Polar Regions, must there be condens'd and fall in snow.

... If such an Operation of Nature were really performed, would it not give all the Appearances of an AURORA BOREALIS? ...

(From *The Writings of Benjamin Franklin*, 1778)

Sophus Tromholt (1851-1896)

In his famous book, *Under the Rays of the Aurora Borealis* (1885), Tromholt devoted a chapter to the description of the aurora in great detail, the distribution, the seasonal variations, magnetic effects, and other aspects. As to the cause of the aurora, he began by stating:

> … It will hardly be of interest to the reader to detail all the theories which have been advanced for explaining the origin and nature of the Aurora Borealis, particularly as not a single one of them has succeeded in being generally accepted, and in giving an explanation of the various peculiarities which observation and research have proved to be part of the phenomenon. Apart from the ancient theories, which may now be considered of little importance, two have been advanced in modern times which may be mentioned. The one is the so-called cosmic theory, according to which the Aurora Borealis should be produced by the Earth's entering into clouds of ferric dust during its passage through space, the molecules of which would, under influence of the terrestrial magnetism, gather in certain conformations, producing the various forms of the aurora

Nineteenth-century explorer Bayard Taylor likened the aurora's rays to "skirts of glorious drapery." Scientists and auroral researchers use many similes such as this to describe the northern lights. (Tom Walker)

and its position in space. This theory appears, however, to have no other adherents than its originators.

The other theory is that interpreting the Aurora Borealis as an electrical phenomenon, and it cannot be denied that the resemblance between the electric discharge in a chamber of rarified air and certain forms of the aurora is so striking that it is impossible to assert that no relation exists between the two phenomena. The electrical theory which seems to deserve most adherence is that advanced by Professor Edlund, of Stockholm. He refers the Aurora Borealis to that class of electrical phenomena which have been denominated unipolar induction. …

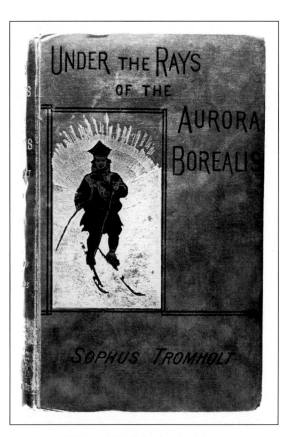

… I have concluded my task of describing the Aurora Borealis, of informing the reader what we know and what we do not know of the phenomenon. Will man ever decipher the characters which the Aurora Borealis draws in fire on the dark sky? Will his eye ever penetrate the mysteries of Creation which are hidden behind this dazzling drapery of colour and light? Who will venture to answer! Only the Future knows the reply. But, nevertheless the student toils yard by yard along the fatiguing road of research.

In concluding the chapter, Tromholt stated:

Lovely celestial display! Before your fascinating mysterious play, in which the enigmatic forces of Nature flood the heavens with light and colour throughout the long Polar night, the golden sunsets of the Pacific Ocean, the gorgeous flora of the Tropics, the resplendent luster of the gems of Golconda, must pale. Lovely celestial display!

Sometimes the colors of the aurora appear mixed rather than in distinct bands such as these rays fanning out over the Chugach Mountains. Edward S. Ellis, in Among the Esquimaux; or Adventures Under the Arctic Circle *(1894), wrote about "innumerable shades, combinations, and mingling of colors" of auroral lights. (Cary Anderson)*

Thomas W. Knox (1835-1896)

Even as early as the nineteenth century some people were quite certain that the aurora was an electrical phenomenon. Thomas W. Knox wrote tongue-in-cheek in *The Voyage of the Vivian* (1884) about the possible use of auroral electricity:

When we have time to spare we will set about devising a machine whereby the electricity of the aurora borealis may be harnessed, and made to do duty in a practical way. We will make it run the dynamos to supply our houses and streets with electric light; it shall propel our machinery, and thus take the place of steam; it shall be used for forcing our gardens, in the way that electricity is supposed to make plants grow; and it shall develop the brains of our statesmen and legislators, to make them wiser and better and of more practical use than they are at present. Hens shall lay more eggs, cows must give cream in place of milk, trees shall bear fruit of gold or silver, teardrops shall be diamonds, and the rocks of the fields shall become alabaster or amber. Wonderful things will be done when we get the electricity of the aurora under our control.

'Yes' responded Fred, 'babies shall be taken from the nursery and reared on electricity, which will be far more nutritious than their ordinary food. When the world is filled with giants nourished from the aurora, the ordinary mortal will tremble. We'll think it over, and see what we can do.'

Karl Selim Lemström (1838-1904)

Karl Selim Lemström, a Finnish

As scientists gathered information about the aurora and at one point thought it was an electrical phenomenon, some talked about harnessing its power for human consumption. Thomas W. Knox jokingly refers to "Fred's Electric Nursery," a nursery where babies become giants through auroral nutrition. (In Thomas W. Knox's The Voyage of the Vivian, *1884, courtesy of Elmer E. Rasmuson Library, University of Alaska Fairbanks; from S.-I. Akasofu)*

scientist, was one of the first to attempt to simulate auroral phenomena in a laboratory. He was strongly convinced that the aurora arises from an electrical-discharge process. He devised an apparatus in which a series of discharge tubes represented the upper atmosphere and an iron ball below represented Earth. Lemström connected one terminal of a high-voltage generator to the discharge tubes and the other terminal to the iron ball. In a demonstration of his apparatus in London in 1879, a glow was seen in the tubes as Lemström predicted. However, he did not have any idea what entity could possibly be the generator for the natural aurora. ■

 # Pioneers of Modern Auroral Science

Kristian Birkeland (1867-1917)

Norwegian physicist Kristian Birkeland was greatly interested in science from his boyhood days. Olav Devik wrote in *Kristian Birkeland as I Knew Him* (1968):

> Kristian Birkeland once told his friend Saeland — later professor of physics — that the first thing he bought with money he had earned himself as a boy, was a magnet. …
>
> … Birkeland started experiments with cathode rays and in a paper from the same year he gives a detailed review of how cathode rays are deflected and sucked in toward a magnetic pole. For the first time he formulated the theory that the sun is sending out cathode rays, mainly from the sunspots.

The modern era of auroral science was opened by Kristian Birkeland. He was an avid student of the aurora and made extensive observations in northern Norway, setting up observatories under the most difficult conditions. The height determin-ation of the aurora was one of his important projects. He also studied in depth the intense electric current of more than 1,000,000 amperes that flows along the auroral curtain. This electric current causes intense magnetic disturbances on the ground. Birkeland envisaged the Sun as the source of the electric current and he theorized that the current is carried by energetic electrons. To demonstrate his idea, he constructed a large vacuum box and studied discharge phenomena around a magnetized iron ball painted with a fluorescent material that was hung from the top of the box. He was able to show that under certain conditions a ring of luminosity appears around the pole, indi-cating that the electron beam "hits" Earth along a ring-shaped area. Thus, Birkeland thought that he had succeeded in repro-ducing the auroral belt. Actually, auroral phenomena are much more complicated than he considered. Birkeland's main inter-est was the behavior of an electron beam in the vicinity of Earth, and he was not really concerned with theories of how the electric current could be generated in the Sun.

Carl Störmer (1874-1955)

Carl Störmer described himself and his view of the aurora:

> In my youth I was a pure mathemati-cian. I never gave a thought to the aurora or similar phenomena. However, I saw Birkeland's most fascinating experiments with the terrella [an iron ball]. These experiments provided the idea of a fundamental problem. At a distance sufficiently great, the magnetic field of the Earth is very much like a dipole [a pair of equal and opposite electric charges]. If Birkeland's analysis was correct, the basic problem was to find the trajectories of the electrically charged corpuscles in the field of a dipole. If

FACING PAGE: *Kristian Birkeland, left, and his assistant Olav Devik pose in their laboratory with a large vacuum box in which Birkeland attempted to reproduce the aurora on an iron ball called a terrella. (Courtesy of University of Oslo; from S.-I. Akasofu)*

Auroras vary in brightness from as faint as the Milky Way to dazzling lights. Norwegian physicist Kristian Birkeland opened the modern era of auroral science by researching auroral height and energy in Earth's atmosphere. (Greg Syverson)

and rays, and; (d) (what was very fascinating) that these very fine phenomena of arcs could stretch (as has been observed) over the heavens as very thin formations some thousands of kilometers long. These conclusions were given directly by the mathematical theory.

(From *Proceedings on the Conference on Auroral Physics*, edited by N.C. Gerson et al., 1954.)

Another important contribution by Störmer was an accurate determination of the height of the aurora. He introduced a photographic triangulation method for this purpose. Sydney Chapman, in the Royal Society's *Biographical Memoirs of Fellows of the Royal Society* (1955) described how Störmer obtained his photographic talent:

In 1930 he told me how he developed into a photographer. When he was a young man at Oslo University he fell in love with a lady whom he did not know and with whom he was too bashful to become acquainted. Wishing at least to have a picture of her, he decided that this was possible only by taking a photograph of her himself, without her knowing. So he procured a small camera that could be concealed under his coat, the exposure

many particles are considered, the problem becomes very difficult. It seemed best to examine first the simpler case where only a single particle and a single stationary dipole were involved. I found this problem so interesting mathematically that I continued with it. At first I thought to conclude the investigation and never repeat it again, but the problem became too interesting.

Although the exact computations (for finding the trajectory of an electric particle in a dipole field) are rather difficult, a relatively simple method permits a simplification. Three differential equations of the second order are involved:

they have two first integrals, implying many interesting things. (These equations have also been of great importance in the theory of cosmic rays which were discovered many years later.) Applying the theory, it was found that a great many phenomena could be explained. The terrella experiments were clarified by the shape of the calculated trajectories of the electrified particles. Applied to aurorae, the corpuscular theory showed: (a) how the auroral particles enter the atmosphere on the night side of the Earth; (b) that the auroral zones can be concentrated in belts around the two geomagnetic poles; (c) that the aurora can be formed in arcs

being made through his buttonhole. He succeeded in this difficult task. The love affair came to nothing; later the lady left Oslo for America and married. Years later she returned to Oslo. By this time he was famous. Meeting her, he told her of the incident. Later he applied his skill in photography (unsuspected by the subject) to obtain pictures also of many of the celebrities of Oslo of those days. Long afterwards, when he was nearing the age of 70, these photographs formed the subject of an exhibition in Oslo. He published illustrated accounts of it (totaling over 60 pages) in 1942 and 1943 — under the title (translated): 'Carl Johan Störmer (Street) snapshots of famous people of the last fifty years.'

Störmer summarized his life work in his book *The Polar Aurora* (1955), which he dedicated to his wife: "To my wife Ada, who never ceased to encourage me to work hard till this book was safely finished."

Sydney Chapman (1888-1970)

Sydney Chapman described his boyhood (from *Chapman, Eighty, from His Friends*, 1968, edited by S.I. Akasofu, et al.):

When I was 14 my father wondered what I should become, and he took me first of all to a builder's merchant, who said that plumbing was quite a good trade. I recall a story of an American plumber who had an only daughter of whom he was very fond and proud. He sent her to college and she got a degree. And then she got a secretarial job. He said, "Yes, she gets $3,000 a year. Not bad for an educated person, is it?" Perhaps I might have been rich if I'd emigrated to America and become a plumber. But no such thought ever crossed our minds.

And then my father took me to an engineer, a man who with his brother had built up a very successful gas engine manufacturing firm. He had been to Manchester University and taken a degree, and was science-minded. He said he would take me into his works at that age, 14, but he said it would be better if I went for two years to a technical school, and I might even, from there, go on to the University. But at any rate, he said he would take me if I wanted to go.

My father took his advice, and I went to a technical institute a few miles away.

… This chemist — a Scotsman — took a kindly interest in me, and suggested I should sit for a county scholarship examination in hope of going to Manchester University.

… The county of Lancashire in which I lived offered 15 university scholarships.

… I took the examinations and I was 15th on the list. At any rate, being 15th

Sydney Chapman, a pioneer in solar-terrestrial physics, established a firm foundation for modern auroral science in 1934 by theorizing the interaction between solar gas flow and Earth's magnetic field. (Courtesy of Geophysical Institute, University of Alaska; from S.-I. Akasofu)

on the list, I was in. At the age of 16, I went to Manchester University to continue the studies in engineering. …

Chapman, with his graduate student Vincenzo C.A. Ferraro, theorized the interaction between solar gas flow and Earth's magnetic field in 1934 and put a firm foundation under modern auroral science. After his retirement from Oxford University in 1953, Chapman joined the staff of the University of Alaska's Geophysical Institute as the scientific advisory director. Not long after, he was

named president of the International Geophysical Year (IGY) and led that remarkable cooperative effort to a great success.

Hannes Alfvén (1908-1995)

Hannes Alfvén, a professor of the Royal Institute of Technology, Stockholm, Sweden, was awarded the Nobel Prize in Physics in 1970 for his fundamental contribution to the physics of ionized gas. He was one of the first to recognize the importance of the magnetic field carried by the solar gas flow.

Alfvén was most appreciated for the originality of his ideas as a pioneer in the field of cosmic electrodynamics, which is fundamental in auroral physics. Often his ideas were too advanced to be appreciated at the time he proposed them. Therefore, he had to work hard to convey his ideas. For example, he was one of the first to propose that the magnetic equatorial plane of the Sun is not flat, but has a wavy structure. In his talk, he showed a slide of a spinning ballerina and pointed out that the solar magnetic equatorial plane is like her fanned skirt. His basic principle in guiding his students and colleagues was that the behavior of ionized gas should be studied first in a laboratory, because ionized gas

Enticing Anchorage city-dwellers out of their warm houses during a cold spell in November 2001, a magnificent red aurora delighted viewers for hundreds of miles across Alaska. (Cary Anderson)

James A. Van Allen, for which the Van Allen radiation belts are named, gives a farewell kiss to his Geiger counter before it is placed on a U.S. satellite. Radiation belts around Earth limit the life of satellites and can be hazardous for astronauts working outside a spacecraft; in January 2002, a rocket was launched from Poker Flat Research Range, near Fairbanks, as part of an experiment using Global Positioning System (GPS) radio signals to understand more about how oxygen emitted from the aurora triggers the expansion of the ionosphere into space, causing radiation belts to form. (Courtesy of James A. Van Allen; from S.-I. Akasofu)

tends to defy theoretical predictions. His motto was "The way to the Cosmos is through the laboratory."

James A. Van Allen (b. 1914)

James A. Van Allen, discoverer of the Van Allen radiation belts, was a pioneer in space science. Before his discovery of the radiation belts he had been greatly inspired by the mystery of electrons in the aurora. He made a number of observations using rockets and detected a stream of electrons in the aurora. His search for the source of auroral electrons led him to discover the Van Allen belts using the first U.S. satellite. With his discovery, many auroral physicists thought that the mystery of the aurora was solved. However, subsequent studies have shown that the origin of the Van Allen belts is related to cosmic rays. It is interesting to note that auroral research led the beginnings of the present space age. ◼

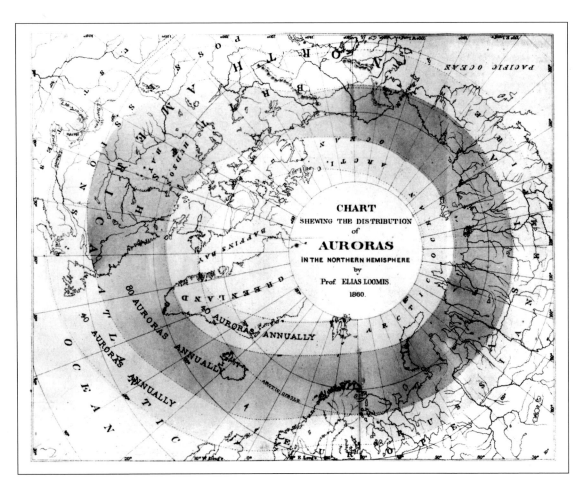

Seeing Probability

The question of how many nights the aurora can be seen in a year at different locations on Earth has been examined by several auroral scientists. In 1860, after a painstaking compilation of auroral records from the past, Elias Loomis, a professor at Yale University, produced a map that shows three belts. In one, the aurora is seen 80 nights per year. The other two belts, representing 40 nights, are located just north and south of the belt of 80

LEFT: *In 1860 Elias Loomis produced the first chart showing the average annual frequency of auroral occurrence. His research helped lead later scientists to define the auroral zone. (In* American Journal of Science and Arts, *1860, courtesy of Yale University; from S.-I. Akasofu)*

FACING PAGE: *Auroras are often described as "ribbons of light" that twist and fold around each other. This display was photographed in Gates of the Arctic National Park in the central Brooks Range. (Hugh S. Rose)*

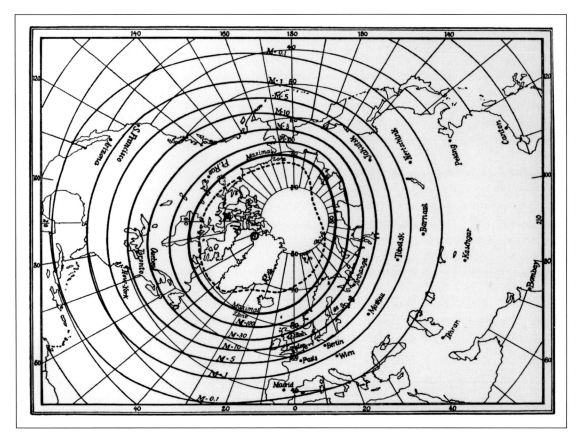

This chart, constructed by Hermann Fritz, shows the average annual frequency of auroral occurrence in terms of number of nights per year. For example, northern Alaska (near top center of map) lies in Fritz's maximal zone where the aurora can be viewed 243 nights per year (he based this number on a long-term average), while viewers in London (near bottom center of map) could see the aurora between five and 10 nights per year. (Courtesy of Hermann Fritz, Das Polarlicht, 1881; from S.-I. Akasofu)

latitudes. Robert E. Peary, during his quest of the North Pole, observed this tendency, providing credibility that he was attaining a very high latitude:

> Nature herself participated in our Christmas celebrations by providing an aurora of considerable brilliancy. While the races on the ice-foot were in progress, the northern sky was filled with streamers and lances of pale white light. These phenomena of the northern sky are not, contrary to the common belief, especially frequent in these most northerly latitudes. It is always a pity to destroy a pleasant popular illusion; but I have seen auroras of a greater beauty in Maine than I have ever seen beyond the Arctic Circle.
> (From *The North Pole*, 1910)

This same fact was noted by several other early polar explorers, and it led some to conclude (erroneously) that the aurora was prone to appear near the edge of the polar pack ice.

nights. It is amazing how accurate many early auroral investigations and publications were because the data available in those times would have been scanty by modern standards. Loomis's work is a good example of this kind, even though his map has since been greatly refined.

In the map produced by Hermann Fritz in 1874, he connected by curves locations where the aurora can be seen equal numbers of nights per year. The number in each curve indicates the number of nights per year in which the aurora can be seen: "0.1" means one night in 10 years; "1" means one night per year; "243" means

243 nights per year. He called these connecting curves "isochasms." A narrow belt centered around this "243" curve is called the auroral zone. These numbers are based on a long-term average. Since the appearance of the aurora depends greatly on the sunspot cycle (a period of about 11 years), the numbers are generally higher than those given in the map during the period when the sunspot number is high.

The maps constructed by both Loomis and Fritz, and later refined by E.H. Vestine, indicate that the occurrence of the aurora reaches maximum in the auroral zone and then decreases toward higher and lower

Auroral Distribution

The auroral zone is defined as the belt around the northern polar region where the aurora can be seen the maximum number of nights per year, about 200 to 250. However, this belt is different from the belt in which the aurora lies at a particular time when the polar region is seen from space well above the North Pole. Throughout the IGY from 1957 to 1958, during which scientists from around the world studied Earth, researchers attempted to determine the actual belt in which the aurora lies. The aurora, however, is a large-scale phenomenon, and it was not possible to study it simultaneously all over the polar region in those days. Since such coordination was not possible in the early days, one of the major unanswered questions during the first half of the twentieth century was where the aurora is distributed in the polar region, as seen if one looked down on Earth from a great distance above the North Pole. It was generally and tacitly believed that the aurora lay along the auroral zone determined by Loomis and Fritz.

Such a study requires recording of the aurora; the first recording was a sketch by perhaps freezing hands. Photography was introduced into auroral study at the end of the nineteenth century. The difficulty faced

Auroral photography has advanced significantly since this photo, one of the earliest successful images, was taken in 1892 by M. Brendel. (Courtesy of U.S. Air Force Geophysical Laboratory Library; from S.-I. Akasofu)

by scientists at that time is well expressed by Sophus Tromholt:

Every attempt I made to photograph the Aurora Borealis failed ... in spite of using the most sensitive dry plates, and exposing them from four to seven minutes. I did not succeed in obtaining even the very faint trace of a negative.
(From *Under the Rays of the Aurora Borealis*)

During the 1957-1958 IGY, auroral scientists made an all-out effort to deter-mine precisely the distribution. For this purpose, they devised a camera that is capable of photographing the entire sky in a single frame, called an "all-sky camera." At that time, a fast fish-eye lens was not available, so it was necessary to construct a camera system consisting of a convex mirror, which reflects any object at any point in the sky from horizon to horizon, and a flat mirror above the convex one, which reflects any image on the convex mirror. The camera is located in a box below the convex mirror and takes a photograph of the image reflected on the flat mirror

BELOW: *This image of the northern auroral oval was captured by the Dynamic Explorer satellite. With the daylight side of Earth at left, the oval appears at top center showing that the auroral oval is widest on the dark side of Earth. (Courtesy of Lou Frank, University of Iowa; from S.-I. Akasofu)*

BELOW, RIGHT: *Outlines of North America, Greenland, and Asia are superimposed onto an image of the northern auroral oval taken by the Dynamic Explorer satellite. "N" indicates the geographic North Pole, "M" the geomagnetic pole. (Courtesy of Lou Frank and John Craven, University of Iowa; from S.-I. Akasofu)*

through a small hole at the top of the convex mirror.

More than 100 of these cameras were installed in the Arctic and Antarctic during the IGY to photograph auroras, one per minute. The all-sky films were analyzed by a number of scientists, including Yasha I. Feldstein and O.V. Khorosheva of the University of Moscow. Feldstein and Khorosheva discovered in 1963 that auroras are distributed along a narrow band encircling the pole, called the auroral oval, which is quite different from the auroral zone.

The presence of the auroral oval was

later confirmed by images taken from satellites. There is another auroral oval over the Antarctic in the Southern Hemisphere. Therefore, Earth has two beautiful rings of the aurora.

Auroral scientists use polar maps, in which the center is the geomagnetic pole, not the geographic pole. As mentioned earlier, Earth's magnetic field at a distance of several Earth's radii can be represented by a hypothetical bar magnet at the planet's center. When the line going through the magnet is extended to Earth's surface, the penetrating point of the line is called the geomagnetic pole. In the Northern

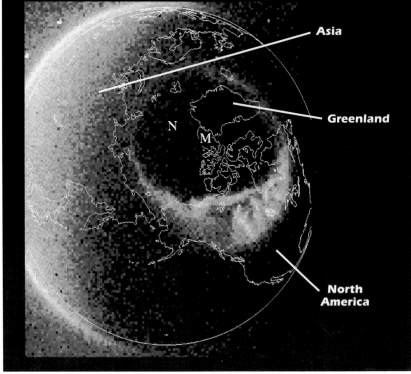

Asia

Greenland

N   M

North America

Hemisphere, the geomagnetic pole is located at Ellesmere Island; in the Southern Hemisphere it is at Vostok in Antarctica. Using such a map, the line from the geomagnetic pole to the Sun is called the magnetic noon meridian, also called the 12 magnetic local time meridian. Extending the line to the nightside, it is called the magnetic midnight meridian. In Fairbanks, the magnetic midnight is about 1 a.m. Perpendicular to the magnetic noon-midnight meridian is the 06 hour magnetic meridian and the 18 hour magnetic meridian. Using these reference points, it is possible to define a new coordinate system for Earth's magnetic field. The geomagnetic poles replace the geographic poles and thus the north geomagnetic pole is at +90° in geomagnetic latitude. Again, geomagnetic latitude is different from geographic latitude. Since the northern geomagnetic pole is located toward the North American continent from the northern geographic pole by about 11.5°, a point in the North American continent is about 11.5° higher in geomagnetic latitude than at a point of the same geographic latitude in Europe.

When the auroral zone is projected on

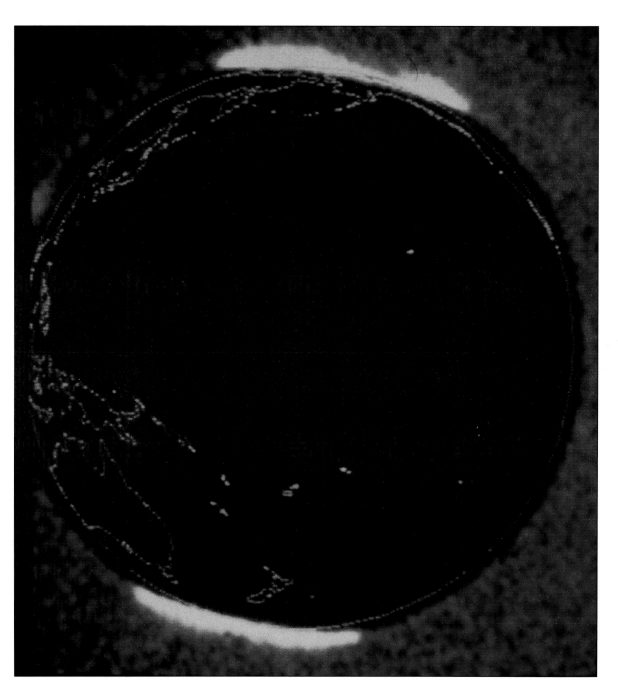

Both the southern and northern auroral ovals are visible in this satellite photo taken from over the Pacific Ocean. Australia is indicated in the lower left portion of the globe by lines superimposed on the image. Alaska appears at the top, just right of center, with the Aleutian Islands dropping into the dark area indicating the Pacific Ocean. (Courtesy of Lou Frank, University of Iowa; from S.-I. Akasofu)

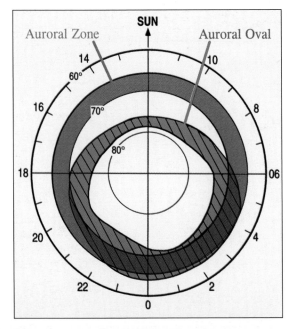

LEFT: *The auroral zone, a narrow band near the poles where displays can be seen about two-thirds of the year (the geographic locations where the seeing probability is highest), and the auroral oval, a narrow band that encircles the poles along which the aurora is actually distributed, are shown here in geomagnetic latitude-magnetic time coordinates. (S.-I. Akasofu)*

BELOW, LEFT: *The southern auroral oval hovers over Antarctica in this image. (Courtesy of Lou Frank, University of Iowa; from S.-I. Akasofu)*

BELOW: *Because the geographic and geomagnetic poles do not coincide, the relative distance between the auroral oval and the observer's field of view (small circles) changes at different local times. Small circles indicate an approximate field of view at geomagnetic latitude 65°. (S.-I. Akasofu)*

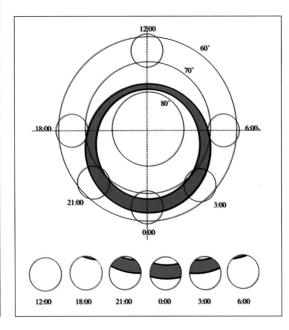

magnetic latitude-local time coordinates, it is seen as a circular belt around the geomagnetic pole. On the other hand, the auroral oval has an oval shape and its center is displaced by about 3° toward the midnight sector; on average, the mid-dayside of the oval is located at magnetic latitude of 76°, but the midnight side is located at 67°.

The aurora in the northern oval is called the aurora borealis, while the aurora in the southern oval is called the aurora australis. The auroral borealis is commonly called the northern lights. The auroras in both the northern and southern ovals are almost a mirror image. Two heavily instrumented jet planes also investigated the question of whether the auroras in the Northern and Southern Hemispheres are the same or in some way different. In 1967, scientists from the Geophysical Institute, University of Alaska and from the Los Alamos Scientific Laboratory, one group flying over Alaska and the other flying simultaneously well south of New Zealand, proved that the aurora borealis and the aurora australis are essentially the same. They develop similar activities and move in a similar way at the same time.

The auroral oval is fixed with respect to the Sun and Earth rotates under it once a day. Thus, the geographic pattern under the oval changes as Earth rotates. Unlike the auroral zone, the auroral oval is not fixed at a particular geographic location.

As Earth rotates under the auroral oval once a day, the distance between the auroral oval and a point of geomagnetic latitude, e.g., 65°, varies. At 12 magnetic local time (about 1 P.M. in Fairbanks), the distance from the auroral oval is greatest. Even if Alaska were in darkness, the auroral oval is too far away to see, more than 600 miles (this was proven during a solar eclipse). As Fairbanks rotates under the auroral oval in the afternoon and evening, this distance becomes shorter. Thus, the auroral oval appears to advance toward Fairbanks. At magnetic midnight (about 1 A.M.), the auroral oval is located right over-head. As morning progresses, the distance begins to increase, so that the oval appears to recede toward the northern horizon.

The radius of the auroral oval varies considerably, depending on auroral activity. When auroral activity is weak, the oval tends to contract to magnetic latitude of 70°, or a little above in the midnight sector. Therefore, one cannot see the aurora from Fairbanks, but it is not that the aurora is absent. When auroral activity is increased, both the northern and southern boundaries of the auroral oval shift equatorward. For moderate activity, it is located about 65° to 67°. The oval expands farther for greater activities to 60°. In such a case, the aurora can be seen overhead at 8 P.M. in Fairbanks. During a great magnetic storm, the oval descends as far as 50°, near Seattle, Washington.

The red emission from excited oxygen atoms (wave length of 630 nanometer [nm]) appears in the upper part of the greenish auroral curtain. Pure oxygen, not the O_2 that people breathe, produces the red light. (Courtesy of Jan Curtis; from S.-I. Akasofu)

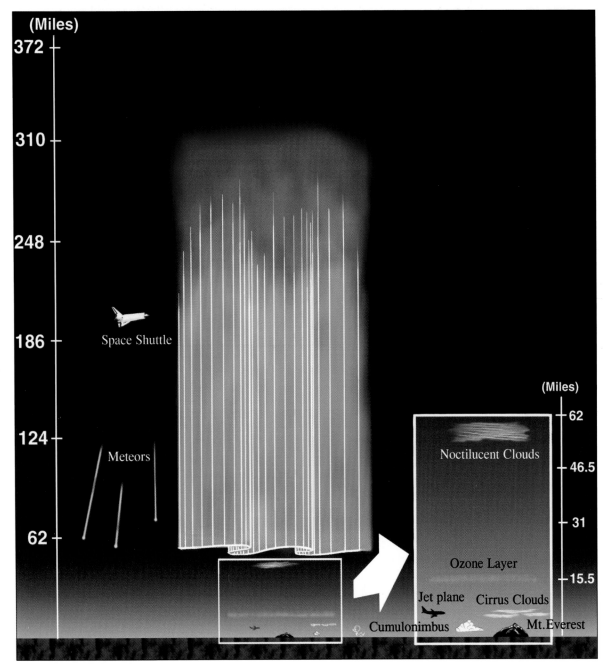

Aurora Forms and Height

A problem that claimed the attention of many early auroral scientists was how to determine the height of the aurora. There was much heated controversy about it, since reported heights ranged from ground level to an altitude of 600 miles. Numerous accounts were published in which observers declared, for example, that the aurora appeared between two houses, or a few thousand yards away, or around the tops of mountains. Even in the *Encyclopedia Britannica* (9th Edition, 1882; 11th Edition, 1910) some of these reports were quoted as reliable. Thus, many scientists sought the cause of the aurora in the lower atmosphere of Earth where clouds are formed. As we shall see, those authors were confused by the effects of perspective.

Frenchman J.J.D. de Mairan and two famous British scientists, Henry Cavendish and John Dalton, accurately determined the height of the aurora to be 50 to 155 miles, while Adam Paulsen, a Danish polar explorer, in 1889 incorrectly reported it to be from 2,000 feet to 43 miles.

The basic auroral form is curtainlike. The auroral curtain appears often in

This illustration shows the height of the aurora compared with the height of Mt. Everest, the cruising altitude of a jet plane, clouds, the ozone layer, meteorites, noctilucent clouds, and a space shuttle. Noctilucent clouds occur 50 miles above Earth and are unique to subarctic latitudes. Space shuttles fly through the aurora, jet planes well below it. (S.-I. Akasofu)

A space shuttle flies through the aurora, the craft's tail pointing to red auroral rays extending into the upper atmosphere. (Courtesy of NASA; from S.-I. Akasofu)

multiples, sometimes as many as five or six, or even more. Its bottom height is almost exactly 60 miles above Earth. The upper height is normally about 200 miles. Thus, a space shuttle flies through the aurora. The bottom height of the aurora is about 10 times as high as the cruising altitude of jet aircraft. One of the reasons why a jet plane flies at an altitude of 40,000 feet is that it flies above clouds, namely above most of the meteorological processes, avoiding turbulent air flow as much as possible. This fact indicates that the aurora is not a meteorological phenomenon. The reason why the aurora appears to be seen on a cold night is because the sky tends to be clear, while the aurora is invisible on a warm night, because it tends to be cloudy; the aurora occurs well above the altitude of clouds. The auroral height is also well above the altitude of the ozone layer. In the polar region, there occurs another beautiful phenomenon, noctilucent clouds, at an altitude of 55 miles, just below the height of the bottom edge of the aurora. The nature of noctilucent clouds is not well established yet.

The auroral curtain is brightest near its bottom and becomes fainter with increasing altitude. When it is quiet, the brightness is uniform in the horizontal direction; such a form is often called a **homogeneous arc**. When the aurora becomes slightly active, it tends to develop vertical striations, called rays, which are actually fine pleats of the auroral curtain. Folded portions of the pleats look brighter than the rest. This was discovered by University of Alaska Geophysical Institute scientist Dr. Tom Hallinan, who used a supersensitive, high-speed television system (ASA equivalent of 1,000,000). The curtain that develops the ray structure is called the **rayed arc**.

Since the basic form of the aurora is curtainlike, imagine looking at a curtain on a theater stage. The look angle depends on the distance of the curtain from the audience seats. This is the same for the aurora. When the aurora is closer, it looks higher in the sky. As the auroral curtain moves closer to an observer or an observer moves closer to the aurora, the arch rises higher and higher. The difference with a theater curtain is that the auroral curtain extends thousands of miles along the auroral oval in the east-west direction, so that the distance toward the eastern and western ends can be as much as several hundred miles. Suppose that the auroral curtain is located at a distance of 100 miles toward the direction of geomagnetic north. Then, for the same reason that electric power line poles look shorter at greater distances, the height near the eastern and western ends looks low. Therefore, the

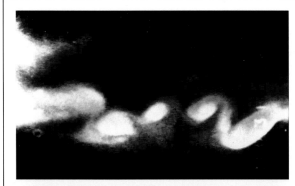

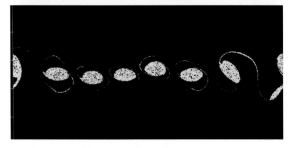

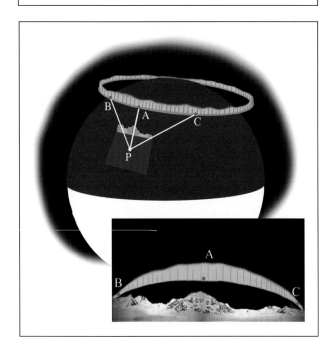

LEFT: This photo, top, taken by the supersensitive, high-speed television system (about 30 frames per second) that Dr. Tom Hallinan used and corresponding illustration, bottom, reveal auroral rays as seen from the bottom of the curtain. Each eddylike structure has a diameter of a few miles and will be seen as a "ray" when the auroral curtain is observed from a distance. A Russian cosmonaut who went through active auroras mentioned that he felt as if he were "passing through magnificent columns of light." The bottom image is a computer simulation by Dr. John Wagner. (Courtesy of Tom Hallinan and John Wagner, Geophysical Institute, University of Alaska; from S.-I. Akasofu)

BELOW, LEFT: An aurora looks like an arch over the horizon because the distance toward the eastern and western ends (B,C) are much greater than the shortest distance (A). The direction of "A" seen from "P" is the direction N (north) of a compass needle. In Fairbanks, it is approximately the direction of Fort Yukon. (S.-I. Akasofu)

auroral curtain looks like an arch above the northern region, when an observer is located south of the aurora. This is also why there were so many reports in which the aurora appeared between two houses and from forests and mountains.

Note that in Fairbanks, Alaska, the direction toward magnetic north is about 30° eastward from the direction toward geographic north. This deviation angle differs from place to place.

The auroral curtain develops folds of various sizes as it becomes active. When the curtain develops a very large fold, it looks like drapery. When a quiet auroral curtain is located near the zenith (directly overhead), the bottom edge looks like a straight line (because one cannot see the side of the curtain) or a slightly wavy line. When a very active auroral curtain with large-scale folds is located near the zenith, it is difficult for anyone to recognize the curtain form, and in fact, it appears to have an entirely different form, called the corona. Again, perspective plays a role here. The observer sees a large number of parallel rays that extend at least a few hundred miles above him; the rays appear to converge near the zenith, just like the distance between a pair of rail tracks appears to converge at a great distance. When the same aurora is viewed simultaneously by an onlooker located a few hundred miles north or south, that observer will see an ordinary curtainlike form. Therefore, contrary to common belief, the corona is not really a different form of the aurora. Its appearance indicates that an observer is almost directly beneath the auroral curtain.

Nature of the Auroral Lights

Another important question is, "What kind of light does the aurora emit?" The answer can tell us two important things. The first is the kinds of atoms and molecules that are emitting light, and the second is why those atoms and molecules emit light. This particular field of science is called spectroscopy, or in our specific case, auroral spectroscopy. The simplest instrument for spectroscopy is a prism; the analyzed pattern of light produced through a prism is called a spectrum.

Until early in the nineteenth century, it was commonly believed that the aurora was caused by the reflection of sunlight from tiny ice crystals in the sky. If that were so, when the auroral light is seen through a prism, it should show the familiar rainbow colors, which range continuously from red

This illustration shows the auroral curtain as seen from different distances: A (zenith), B (meridian distance), and C (far distance); and also the eastern and western views from B. (S.-I. Akasofu)

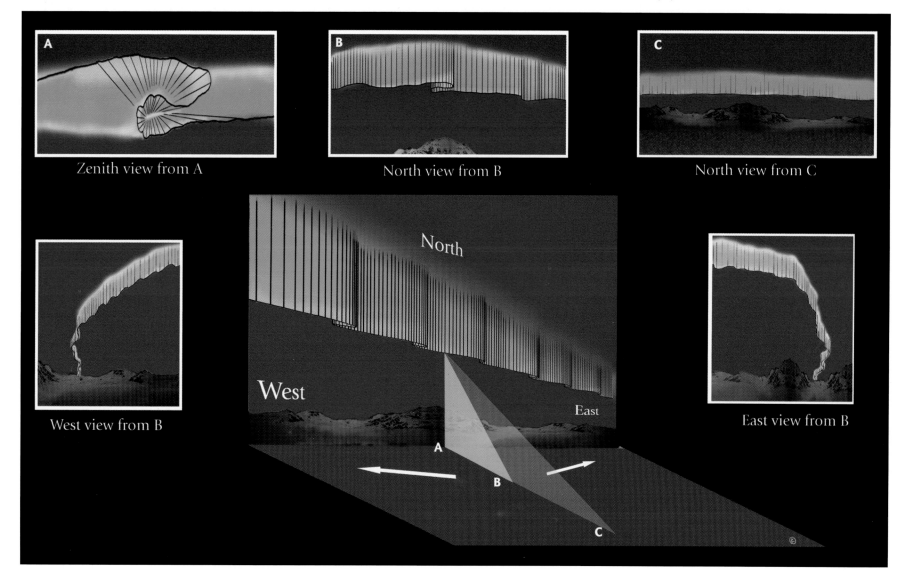

Zenith view from A

North view from B

North view from C

West view from B

East view from B

to violet; such a spectrum is thus called a continuous spectrum.

Norwegian physicist Anders Jonas Angstrom was one of the first to use a prism to study the aurora; he found that auroral light is nothing like that of a rainbow. The auroral spectrum is not continuous and it consists of many lines and bands of different colors with dark spaces among them. The lines are emitted by atoms and the bands by molecules.

In the middle of the nineteenth century, physicists were aware that a spectrum consisting of such lines and bands of light can be produced from a vacuum glass vessel when a high voltage is applied through electrodes inserted

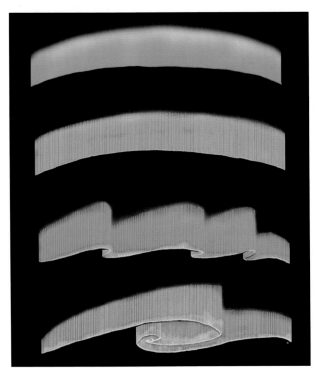

into the vessel. This process is called a high vacuum electrical discharge. A vacuum must be created because standard air is a good insulator, but rarefied air is not. These lights come from the atoms and molecules remaining in the vessel when they are struck by high-speed electrons that are shot from the negative electrode connected to the high-voltage source. The electrons carry the discharge current from the negative electrode to the positive one.

To illustrate this effect, a thin glass tube in which a vacuum has been induced is filled with a small amount of neon gas, and is then connected to a high voltage source. Electrons stream from the negative end to the positive end of the glass tube. The neon atoms are hit by the streaming electrons and their internal state changes. This process is called excitation in physics. However, a neon atom cannot remain in the excited state for very long and must revert to the ground state, or normal condition, within nanoseconds. During this process of returning to ground state the extra energy the atom received from electrons by the collision is released in the form of the familiar bright red light. This particular emission is a characteristic color of neon atoms, and no other kinds of

During quiet times, the auroral curtain has a uniform brightness horizontally. As the aurora becomes slightly active, the curtain develops fine pleats that look like vertical striations, called "rays." As auroral activity increases, folds of larger scale develop; folded portions look brighter than the rest. (S.-I. Akasofu)

atoms can emit the same light. Similarly, when mercury vapor is sealed in the tube and excited, a bright bluish-white light is produced. Again, no other kinds of atoms can emit this same color light. Physicists have compiled a list (or "spectrum") of the lights emitted by different kinds of atoms and molecules, organized by wavelength. In this way scientists can now recognize most of the atoms and molecules that emit auroral lights in the polar upper atmosphere and even in nebulas and galaxies.

Angstrom found that the most common light form of the aurora, a whitish-green color, had a wavelength of 5567 Angstrom (the present accurate determination is 5577 Angstrom [or 557.7 nm]). This line is called the green line by auroral scientists. From the time of Angstrom's discovery (1868), to as late as 1924, however, this green line was a mystery because no one could successfully find an atom that produced it in a vacuum glass vessel.

For example, Sir William Ramsay, in *The Gases of the Atmosphere: History of their Discovery* (1905), noted:

Shortly after the wave-lengths of the lines in the spectrum of krypton were published, Sir William Huggins, in a private letter, suggested to Sir William Ramsay that its brilliant green line appeared to be identical with that seen in the spectrum of the aurora borealis. The same remark was made somewhat later by Professor Schuster, in a letter to Nature.

The aurora borealis or Northern

Since the bottom edge of auroral curtains is known to be about 60 miles from Earth, it is possible to infer the distance between the curtain and the observer by measuring the elevation angle. If the elevation angle happens to be 45° in Fairbanks, the aurora is located about halfway between Fairbanks and Fort Yukon. (S.-I. Akasofu)

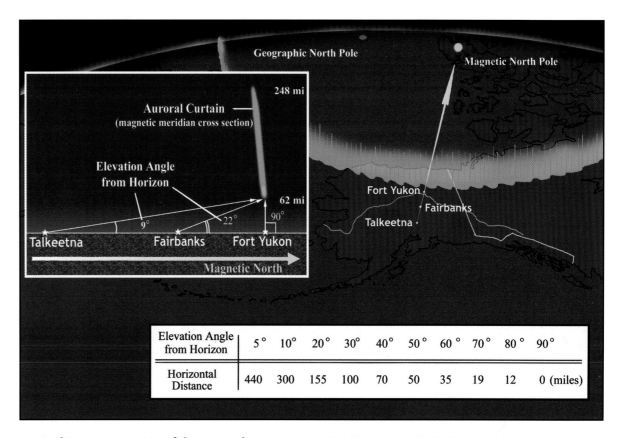

Elevation Angle from Horizon	5°	10°	20°	30°	40°	50°	60°	70°	80°	90°
Horizontal Distance	440	300	155	100	70	50	35	19	12	0 (miles)

Lights generally appears in the north, on frosty evenings, as a luminous arch, from which streamers descend, and emit light, sometimes white, sometimes green, and sometimes crimson. The height of this arch appears to be from 50 to 125 miles. The spectrum contains numerous lines, all of which have been shown … to be identical with strong lines in the spectrum of krypton, but the strongest is one of wave-length 5570 Angstrom units.

Now this krypton line persists at great rarefactions. Even when the amount of krypton is reduced to one twenty-three-millionth part of its normal pressure, the line still is visible. It can be calculated that the pressure of the atmosphere would be equal to that amount at a height of 80 miles, a number which falls within the limits given above.

Sir William Ramsay has succeeded in producing an artificial aurora by causing a ring-shaped discharge to take place through krypton in the interior of a flask, and by a powerful electro-magnet, suitably placed, the "streamers" can also be reproduced. Such an aurora shows all the peculiarities of the natural aurora, including the spectrum characteristic of krypton.

In his great treatise of the auroral spectra entitled *Aurorae; Their Characters and Spectra*, John Rand Capron concluded in 1879:

As the general result of spectrum work on the Aurora up to the present time, we seem to have quite failed in finding any spectrum which, as to position, intensity, and general character lines, well coincides with that of the Aurora. Indeed, we may say we do not find any spectrum so nearly allied to portions even of the Aurora-spectrum, as to lead us to conclude that we have discovered the true nature of one spectrum of the Aurora (supposing it to comprise, as some consider, two or more). The whole subject may be characterized as still a scientific mystery — which, however, we may hope some future observers, armed with spectroscopes of large aperture and low dispersion, but with sufficient means of measurement of line positions, and possibly aided by photography, may help to solve. The singular absence of Aurorae has, for some time past, given no opportunity in

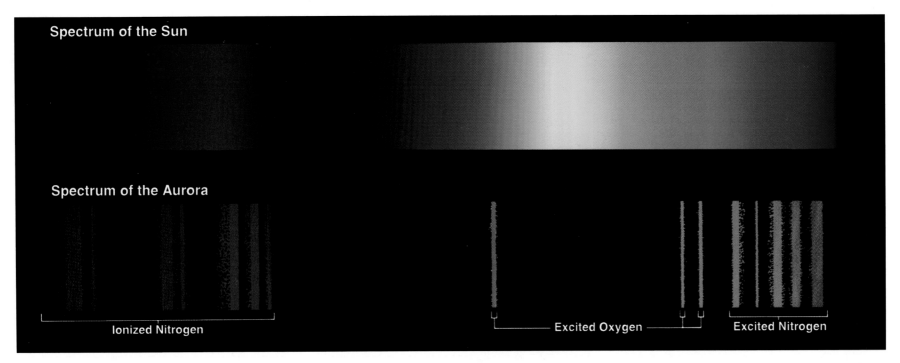

Spectrum of the Sun

Spectrum of the Aurora

Ionized Nitrogen — Excited Oxygen — Excited Nitrogen

The solar spectrum, top, shows the familiar rainbow colors, a continuous transition from red to violet, while the auroral spectrum consists of many lines and bands of different colors, each characteristic of a particular type of molecule. (S.-I. Akasofu)

that direction. May some of my readers be more fortunate in obtaining opportunities of viewing the glorious sky-fires, and assist to unravel so interesting a paradox!

In 1925, two Canadian scientists, J.C. McLennan and G.M. Shrum, finally discovered that this whitish-green light is emitted from **atomic oxygen** (O). In the lower atmosphere, oxygen exists in a molecular form (O_2). However, at the height at which the aurora appears, the oxygen molecules are found separated into their two oxygen atoms. The difficulty associated with this identification was that it takes 0.7 seconds for O atoms to emit the 557.7 nm light after the excitation (most atoms emit their lights within nanoseconds). As a result, collisions with other atoms and molecules take away the excitation energy, preventing O atoms from emitting their light in a poor vacuum tube. Atomic oxygen can also emit a dark red light (wavelength of 630.0 nm) under an ultra-high vacuum condition (above 150 miles in altitude) — the "bloody red color" that caused so much fear in medieval days.

This line is called the red line by auroral scientists.

When the aurora becomes very active, the bottom of the folded curtain is often tinted a crimson red, one of the most beautiful colors of the aurora. This light is a band and is emitted by excited nitrogen molecules (N_2). This emission is called the aurora with the red lower border.

Lars Vegard, a Norwegian scientist, pioneered the field of auroral spectroscopy. During the first half of the twentieth century, auroral light was studied in great detail by scientists from the United States, Canada, Scandinavia, the USSR, and elsewhere.

Auroral spectroscopy tells us that the aurora is an electric "discharge phenomenon" because the auroral lights are emitted by atoms and molecules in the upper atmosphere when they are hit by a beam of high-speed electrons. It is interesting to note that the atmosphere at the auroral altitude corresponds approximately to a high vacuum in an ordinary electronic tube (like the radio tube of pre-transistor days), so that the entire upper atmosphere can be considered to be a discharge tube. As energetic electrons penetrate into the upper atmosphere from above, they collide with nitrogen molecules. The impacting electrons are so energetic that upon a collision they "kick out" or eject an electron from the nitrogen molecule. As a result, a nitrogen molecule loses a negative charge (an electron is negatively charged), and becomes positively charged. This process is called ionization and the product is an ionized nitrogen molecule (N_2). The energetic electron loses only a tiny fraction of its energy by a single collision, so that it continues to collide with other nitrogen molecules, ionizing a large number of them along its way. The ionized nitrogen molecules, meanwhile, emit a strong violet and ultraviolet band (391.4 nm).

The secondary electrons, the ejected electrons, are also energetic. If an oxygen atom is hit by them, it gets excited. As it returns to the ground state, the oxygen atom emits the whitish-green light of wavelength of 557.7 nm.

The bottom height of the auroral curtain is 60 miles in altitude because most auroral electrons lose all their initial energy at that altitude. However, if the primary energetic electrons are energetic enough, they can penetrate down to an altitude of about 55 miles. Much of their energy is lost by the time they get down to such a low altitude, so that they are incapable of ionizing nitrogen molecules. But they can still excite the nitrogen molecules and cause them to emit the crimson red color, producing the aurora with the red lower border.

The aurora emits both ultraviolet and infrared lights and x-rays. Both the ultraviolet and x-rays are absorbed in the atmosphere before reaching the ground. A process related to the aurora, but located well above the auroral height, emits radio waves over a wide frequency range. This radio "noise" can be heard by a radio receiver aboard satellites, but not by humans on the ground. It is quite intense in the broadcast band (500-1600 kHz). Fortunately, the ionosphere is shielding us from this radio noise. Without the ionosphere, we would not be able to use a radio. On the other hand, from outer space Earth would be heard as a very "noisy" planet. Thus, the auroral processes produce electromagnetic waves of all frequencies (or wavelength) from radio waves to x-rays, except for l-rays (gamma).

Anders Jonas Angstrom (1814-1874) was one of the first to discover the auroral spectrum and found that it was quite different from the solar spectrum. (Courtesy of W. Stoffregen; from S.-I. Akasofu)

Does the Aurora Make Sound?

In his book *Under the Rays of the Aurora Borealis*, Sophus Tromholt noted:

There is no point relating to the Aurora Borealis which is more disputed than the sound which some say accompanies the phenomenon, at all events at certain times. It is described as of various natures, viz., cracking, whizzing, and hissing, from nearly every part of the world where the Aurora

FACING PAGE: *This active auroral curtain is approaching the zenith, showing a corona-type display. The lower part of the curtain is emitting the green line (atomic oxygen atoms in a low vacuum condition); the upper part is emitting the red light (atomic oxygen atoms in an ultra-high vacuum condition). (Courtesy of Jack Finch; from S.-I. Akasofu)*

RIGHT: *Bright lights make it difficult to view the aurora from inside the city, but from the Chugach Mountains a green curtain can be seen swirling above Anchorage. (John W. Warden)*

Borealis is visible, and the faith in the "sound" is orthodox among the Eskimo of Greenland and the Lapps of Finmarken and the Tchuctches of New Siberia. Even [members of] the English Circumpolar Expedition to Fort Rae (1882) assert that they distinctly heard the sound one night when passing up the Great Slave Lake; and in latitudes much further south too, people aver having heard the noise. Well, the remarkable part of this question is, that all other scientists, who have sojourned for a length of time in northern regions, have never heard the slightest sound which could with any amount of certainty be ascribed to the Aurora Borealis.

Without absolutely refusing to believe in the possible existence of such a sound, I fancy that there must be some acoustic deception or misunderstanding which has created this belief in an auroral sound. During my stay at Koutokaeino [Norway], I was daily surrounded by people who believed as firmly in the sound as in the Holy Gospel, yea, at Bossekop they even told me that they did not think there was any Aurora Borealis at all until it whizzed, and still I maintain, that of all the intense aurorae I have observed in various parts of the Arctic regions, and which I am sure I have watched with more attention than is generally bestowed on them, every one has been perfectly silent.

Many people have reported hearing a crackling or hissing sound coming from the aurora on occasion. Clarence A. Chant, for instance, reported in 1923 (from *Auroral Audibility*, 1973, by Sam M. Silverman and Tai-Fu Tuan):

We watched this display approaching from the north. At first there was no sound, but as it got nearer, we heard a subdued swishing sound, which grew more distinct as it approached, and was loudest when the ribbon or belt of light was right overhead.

Another observer, D.M. Garber, wrote in 1933 (also from *Auroral Audibility*):

The spectacle was so awe inspiring that a dog team was stopped, and I sat upon the sled for more than an hour absorbing the marvelous beauty of this most unusual display. As we sat upon the sled and the great beams passed directly over our heads, they emitted a distinct audible sound which resembled the crackling of steam escaping from a small jet.

To study such a "phenomenon," auroral scientists must, first of all, record and analyze it. Unfortunately, scientists have not been able to record any sound even with modern, sophisticated audio equipment. Auroral scientists have discovered that an intense auroral display does produce very low frequency pressure pulses (infrasonic), but their frequency is too low to be detected audibly. ■

Nearly vertical rays descend as a motorist stops along a road in Southcentral Alaska to observe. Active rays sometimes look ragged along their edges as they streak and flutter across the sky. (Cary Anderson)

Auroral curtains are not simply hanging in the sky. The aurora often moves so violently, Arctic explorers could not find a way to describe active auroral displays over the entire polar sky. However, before the IGY, there was no systematic coordination of auroral observations in different parts of the polar region to study large-scale auroral displays. A detailed analysis of all-sky films taken from many locations during the IGY revealed that intense global auroral activity tends to be fairly systematic all along the auroral oval with a lifetime of about three

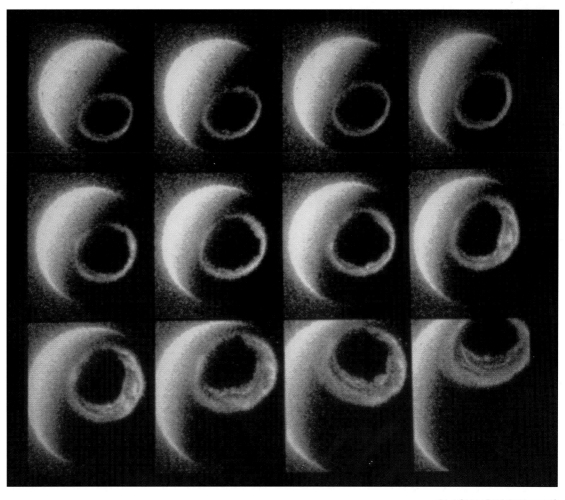

This series of photos taken by the Dynamic Explorer satellite shows the growth and decay of an auroral substorm. The larger half-circle is the side of Earth facing the Sun; the smaller circle is the auroral oval. The time progresses from left to right in each row, about two minutes apart. The aurora in the midnight sector brightens first and the brightness spreads along the auroral oval. Then the aurora in the night sector moves northward, resulting in a bulge. (Courtesy of Lou Frank, University of Iowa; from S.-I. Akasofu)

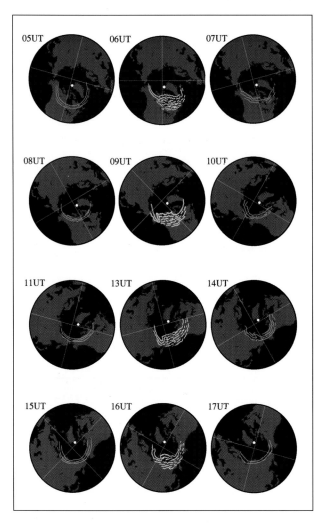

05UT 06UT 07UT
08UT 09UT 10UT
11UT 13UT 14UT
15UT 16UT 17UT

An illustrated view from above Earth shows four auroral substorms occuring successively, beginning at 05, 08, 11, and 15 UT (Universal Time [Greenwich mean time]). Auroral curtains are shown by wavy lines; the geomagnetic pole is indicated by a white dot near the northwestern tip of Greenland. This is a very active night; displays can be seen three times at 09, 13, and 16 UT (about 11 P.M., 3 A.M., and 6 A.M., respectively) in Fairbanks, Alaska. The display is too far north to be seen from there at 06 UT. In a moderately active night, two substorms may be seen. (S.-I. Akasofu)

Auroral Display During a Substorm

Auroral displays during a substorm are different in the evening, midnight, and morning sectors. A substorm begins when an auroral curtain in the midnight sector or the late evening sector becomes suddenly brighter. If there are many curtains, the southernmost one brightens first. After the sudden brightening, the auroral curtain begins to move northward. As a result, the auroral oval bulges in the midnight sector.

The bulging causes a large-scale wavy structure at the western end of the bulge located in the late evening sector. The wave develops into a surge and begins to propagate toward the early evening sector, with a speed of about one mile per second. This particular display is called the westward traveling surge. Within the surge, there occur violent auroral motions. Rays move rapidly along the curtain that develops a variety of folds. The bottom edge of the curtain is tinted by a beautiful pinkish light, the red lower border. When the surge is seen around the zenith, one can observe the most spectacular corona-type displays.

In the morning sector, auroral curtains tend to develop a particular shape of folds that project onto the ground like the inverted Greek character Ω. For this reason, such a fold is called an inverted omega band or just an omega band. The omega-shaped folds propagate eastward toward the morning twilight sky.

Although it has not been mentioned so far, there is another type of aurora. When the curtainlike aurora is quiet, this particular type of aurora looks like the Milky Way, stretching along the curtainlike aurora, but located just south of it. For this reason, this aurora is called the diffuse aurora. The width of the diffuse aurora can be very wide (a few hundred miles). When it covers half of the sky with a uniform brightness, we do not necessarily recognize its presence. Often, we get the impression that faint stars are not visible during a diffuse auroral display. When the diffuse aurora develops dark streaks in such a situation, scientists tend to pay more attention to the dark streaks, rather than to the uniform diffuse glow. Some scientists call them the black aurora.

When a substorm occurs, the diffuse aurora increases its brightness. Further, towards the morning twilight sky, the dark streaks separate the diffuse aurora into many auroral curtains. During an intense substorm, those curtains develop many complicated folds. Since the curtains are rather faint, folds become the prominent

hours, after which it subsides. This particular auroral activity is called the auroral substorm. When auroral activity is weak, the auroral substorm occurs every six hours or so (thus, it can be seen once a night at a location). However, when auroral activity is intense, the auroral substorm occurs almost successively (thus, it can be seen three to four times a night at a location). Such a case is called the geomagnetic storm.

feature and look like isolated bundles of lights. For this reason the auroral substorm used to be called the break-up, meaning the curtainlike form is broken up into many rays. When such auroral curtains are faint, it's difficult to recognize the bottom edge and the surface of the curtain and they appear like a group of cumulus clouds, or patches of light. Further, the brightness of the rays tends to flicker. For this reason,

An auroral corona offers viewers standing directly below an active curtain the same perspective as when viewing railroad tracks that appear to converge at a distant point. (Cary Anderson)

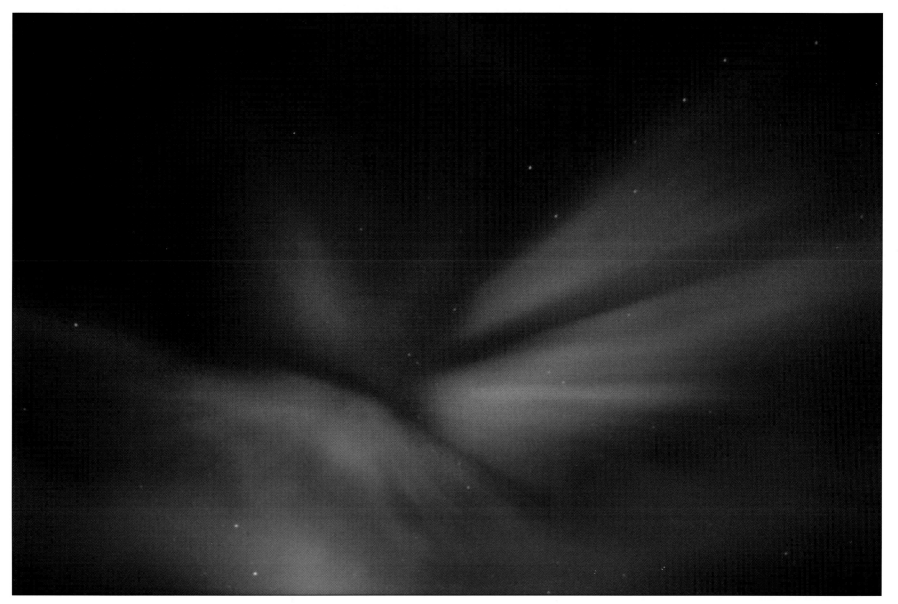

they are called pulsating patches. The patches as a whole drift eastward with a speed of a few miles per second. Patches appear and disappear as they pulsate; a few seconds after the disappearance, they reappear slightly east of the location where they disappeared. Sometimes, the diffuse aurora as a whole deforms and produces a torchlike structure that propagates eastward.

BELOW: *The structure and optics of an all-sky camera allow the widest possible view of the night sky in polar regions. More than 100 of these cameras were operated during the 1957-1958 IGY, revealing auroral activity on a large scale. At* **RIGHT,** *a scientist makes a final check of an all-sky camera during an Arctic sunset. (Both, courtesy of Geophysical Institute, University of Alaska; from S.-I. Akasofu)*

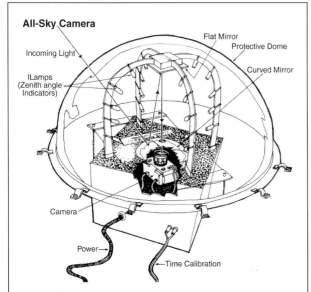

All-Sky Camera
Incoming Light
ILamps (Zenith angle Indicators)
Flat Mirror
Protective Dome
Curved Mirror
Camera
Power
Time Calibration

Why Does the Auroral Curtain Move?

Since we have learned how auroral curtains move during an auroral substorm, it is important to learn further why they move. Perhaps the simplest way to explain auroral motions is to draw the analogy with an oscilloscope, an instrument used by electronic engineers and in hospitals.

In a hospital, medical doctors examine a heartbeat by using an oscilloscope. On the screen of their oscilloscope, a bright dot moves up and down to indicate heartbeats. This is done by converting the heartbeat signals into electrical signals that are applied to a pair of electric plates in the oscilloscope tube. During one half of a heartbeat, the upper plate becomes positive and the lower plate becomes negative. During the other half, the upper plate becomes negative, the lower plate positive. An electron beam is generated at the end of the tube and is shot through space between the two plates. As the electron beam passes the space between the two plates, electrons (carrying a negative charge) are attracted toward the upper plate during one half of the heartbeat, while they are attracted towards the lower plate during the other half. The beam hits the back of the screen, which is coated with a fluorescent material.

As an electron beam moves up and down, the impact point does too. On the screen, a bright point moves up and down, telling doctors how the heart beats.

Our auroral situation is similar to the case of the oscilloscope. The screen of the tube corresponds to the upper atmosphere. When an electron beam hits the upper atmosphere from above, each electron collides with atoms and molecules in the upper atmosphere. When they are hit, atoms or molecules emit their own lights within nanoseconds, except O atoms that take 0.7 seconds. In the case of the aurora, the electron beam must be a sheet beam to generate a curtainlike structure. When a sheet beam of electrons penetrates into the upper atmosphere, the electrons collide with upper atmospheric atoms and molecules within the space occupied by the sheet beam. The aurora appears as a thin sheet because the electrons are constrained to move mostly along the magnetic field lines in the sheet space.

Suppose that the electron sheet beam aligned in the east-west direction is penetrating into the upper atmosphere and is shifting northward. The impact space in the upper atmosphere then shifts northward, and atoms and molecules in the new space emit the auroral light; within a second, the old impact space becomes dark. Just like the case of the oscilloscope tube, this change appears to the observer as if the auroral curtain shifts northward. When the sheet beam develops folds, it produces an auroral curtain with folds.

In an oscilloscope, the electron beam is modulated by an electric field between a

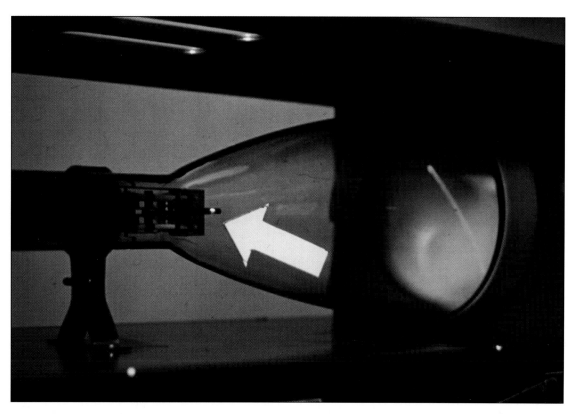

pair of plates and/or a magnetic field produced by a coil. Using this analogy, auroral scientists are trying to learn from auroral displays during an auroral substorm how both electric fields and magnetic fields in space around Earth change. Electric current generates magnetic field. Since the aurora is caused by an electric discharge, the auroral substorm is associated with intense electric currents of more than 1,000,000 amperes, which cause intense magnetic fields, namely magnetic disturbances. When intense substorms occur successively, substorms develop into a storm, called a geomagnetic storm. ▪

In an oscilloscope tube, an electron beam is shot from the cathode toward the back of the screen. As the electron beam goes through between two electric plates, the negatively charged electron beam is deflected toward the positive (upper) plate, so that the impact point moves upward. The back of the screen is coated with a flourescent material and emits light when impacted by the electron beam. If the polarity of the plates is changed, the electron beam is deflected toward the lower plate, so that the impact point moves downward. These upward and downward movements of the impact point are seen as the movements of the bright spot on the screen. (S.-I. Akasofu)

What Kind of Auroral Displays Are You Watching?

Auroral displays are infinitely variable, but viewed in terms of the auroral substorm, some order can be made of the variety. This section allows serious aurora watchers to recognize typical auroral displays. Like bird-watching, aurora-viewing offers opportunities to simply watch and listen or to go a step further and also identify.

In the following pages, auroral displays are illustrated by traditional photographs (showing what you can see with the naked eye), series of all-sky photographs (illustrating how the display develops in time over the entire sky), and satellite images (presenting how it looks when seen from well above Earth). Use this section to help identify auroral displays as you observe them. Images are credited individually; all are from S.-I. Akasofu.

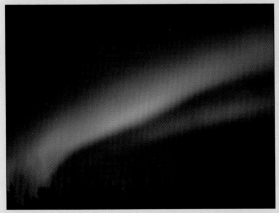

JAN CURTIS ▲

JAN CURTIS ▲ ▼

LEE SNYDER ▲

These photographs show actual examples of the auroral forms illustrated on page 62 (shown below). When an auroral substorm is not in progress, the auroral curtain all along the auroral oval is hanging quietly in the sky, with no structure in the curtain. Such a form is called a homogeneous arc and can be seen most often in evening and midnight skies (upper left). When the aurora becomes a little active, the curtain develops vertical striations; such a form is called a rayed arc (upper center). As auroral activity increases, folds of various scales (10 miles to 100 miles) develop, together with the ray structure (upper and lower right). Can you identify which forms you are watching?

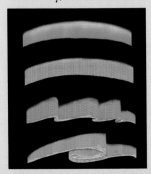

JAN CURTIS ▲ ▶

S.-I. AKASOFU ▼

KATSUHIRO MORI ▼

Compare these photos to the illustrations on page 61. When the aurora is located 100 miles north or south of an aurora watcher, it looks like a curtain or arc. However, when an active aurora is located near the zenith, it looks like a bundle of rays emanating in all directions from a point near the zenith (upper left and right). Such a form is often referred to as the corona. When you watch the curtainlike aurora near the eastern or western horizon, it looks like bright smoke rising from the horizon (lower left and right). You are looking at the aurora several hundred miles away near the horizon. The bottom height of the auroral curtain is about 60 miles, but the perspective effect gives you the impression that the aurora touches the ground.

◀ TAKESHI AND AIKO MATSUO; ▲ JACK FINCH

JAN CURTIS ▼

FACING PAGE: *When you see a greenish-white light (far left), you are watching emission from oxygen atoms as they are hit by high-energy electrons. When you see a dark red light near the upper part of the auroral curtain (upper right), you are watching the emission from atomic oxygen as it is hit by low-energy electrons. The bottom of an active auroral curtain is often tinted by pinkish light (lower right), the emission from nitrogen molecules.*

BELOW: *During typical auroral activity on a global scale, namely the auroral substorm, the aurora shows different characteristics of displays, depending on different local times. It's important to know what time you are watching the aurora. The figure on the left shows schematically different characteristics of auroral displays over the whole polar sky, as if you can watch them from high above the north magnetic pole (top [12] is midday; bottom [0] is midnight; left [18] is evening; and right [06] is morning). The right-hand figure is an actual satellite image of the aurora over the nightside of Earth.*

▼ **LEFT, S.-I. AKASOFU; RIGHT, COURTESY OF THE DEFENSE METEOROLOGICAL SATELLITE PROGRAM**

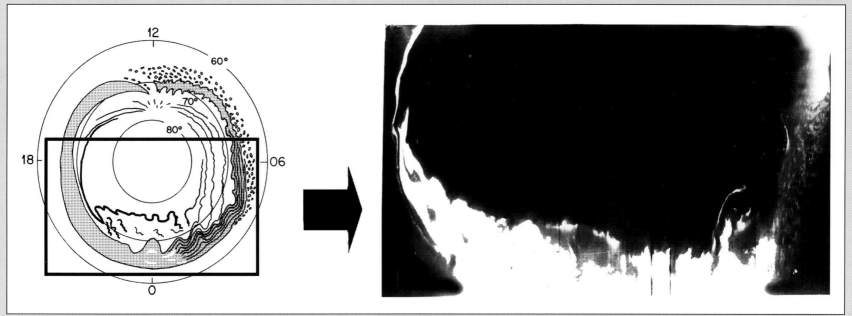

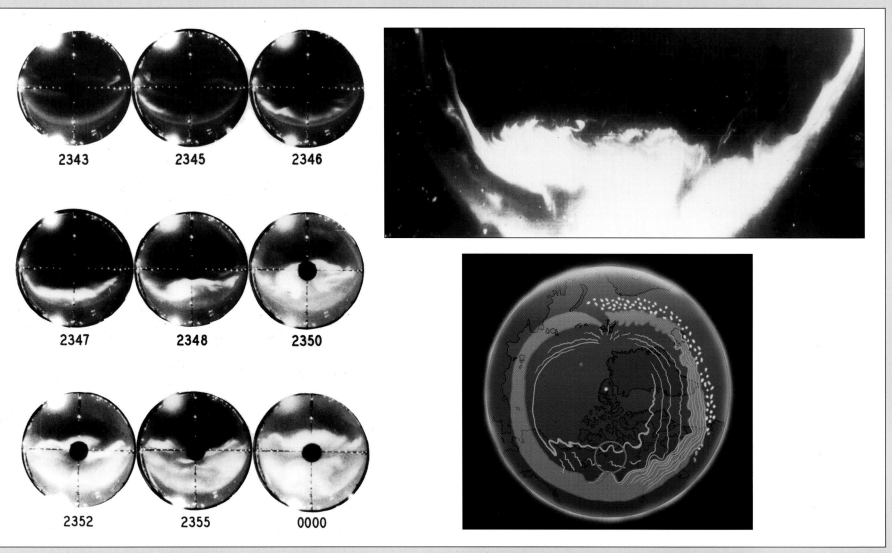

2343 2345 2346

2347 2348 2350

2352 2355 0000

▲ LEFT, COURTESY OF GEOPHYSICAL INSTITUTE, UAF; TOP RIGHT, COURTESY OF DEFENSE METEOROLOGICAL SATELLITE PROGRAM; BOTTOM RIGHT, FROM S.-I. AKASOFU

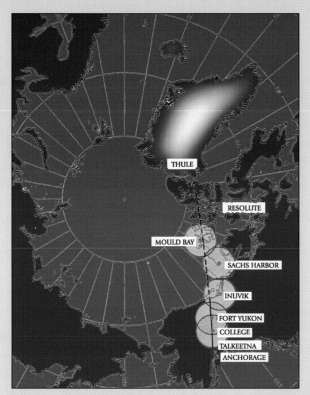

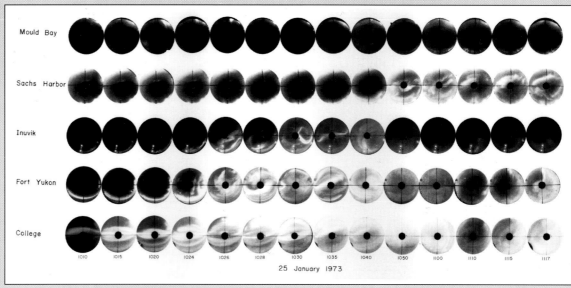

FACING PAGE: *An intense auroral activity, called an auroral substorm, begins when a homogeneous arc or a slightly rayed arc located near the zenith of geomagnetic latitude of 65° (Fairbanks, Alaska; Tromsø, Norway; or Yellowknife, Canada) increases suddenly in brightness and starts to move northward. The figure on the left shows a series of all-sky photographs taken a few minutes apart (the time is in Alaska Standard Time), showing the onset of an auroral substorm. In each circular photograph, the top is oriented toward the magnetic north (the direction of a compass needle); the left side is directed westward and the right side is directed eastward. It is difficult for a single watcher to see the whole pattern of global auroral activity, since he or she can only see a small portion of the whole polar sky; the range of sky for a single watcher is illustrated by a circle in the image at lower right. The northward motion of auroras in the midnight sector results in a large bulge (upper right).*

ABOVE LEFT AND ABOVE: *One way to observe how far the northward-moving aurora can reach is to have a north-south chain of all-sky cameras. The left-hand figure shows the chain operated by the University of Alaska Fairbanks's Geophysical Institute. Each circle shows the field of view of an all-sky camera. From the figure above, you can see that auroral activity that began near the zenith of Fairbanks (College) at 1015 Universal Time (about midnight in Alaska) advanced rapidly all the way to Sachs Harbor over Banks Island, Canada, in about 30 minutes.*

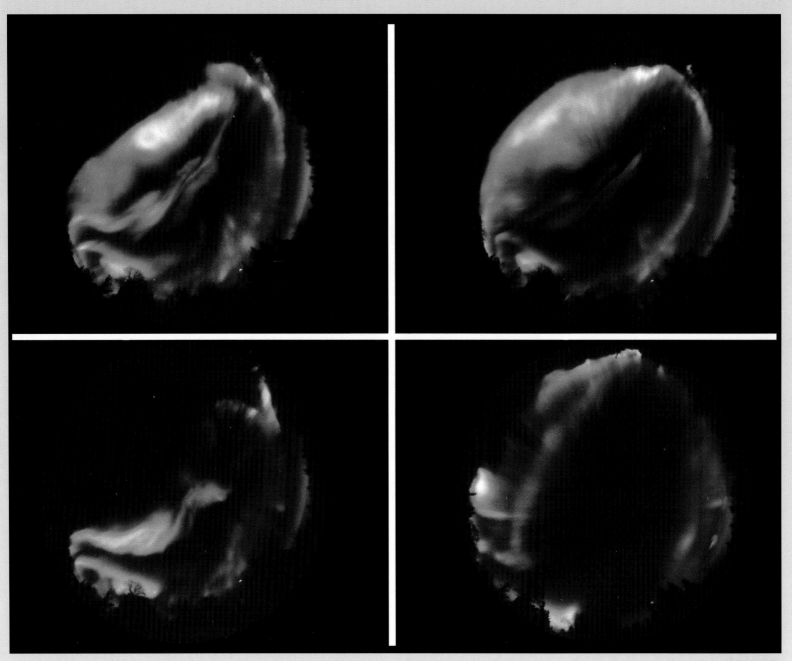

FACING PAGE: *The northward movement of auroral arcs at the beginning of an auroral substorm is captured here by a fish-eye lens. Auroral arcs in the southern sky became suddenly active and started to move northward (clockwise beginning from lower left). Magnetic north is in the direction of the upper left corner of each photograph.*

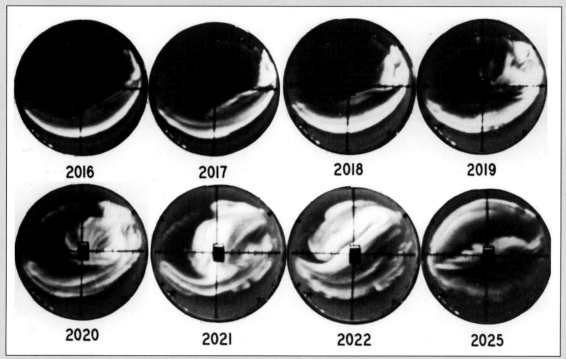

2016 2017 2018 2019

2020 2021 2022 2025

▼ FROM S.-I. AKASOFU; ▲ COURTESY OF GEOPHYSICAL INSTITUTE, UAF

▲ JAN CURTIS

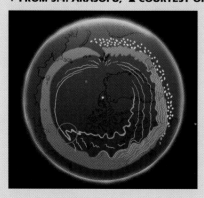

THIS PAGE: *The first sign of an auroral substorm in the evening is an increase in the brightness of the auroral curtain, which propagates from the eastern horizon along the curtain. Rapidly moving structures (folded arcs, shown above, right) near the eastern horizon will move westward with a speed of several miles per second and advance toward the western horizon after passing over your head. This particular activity is called the westward traveling surge. You are watching the western edge of the northward expanding auroral structure in the midnight sector. The figure above, left, is a series of all-sky photographs, showing how the surge moves across the sky (the time is in Alaska Standard Time). The figure at left shows the location of the westward traveling surge, seen from above the north magnetic pole, and the circle indicates the field of view of an observer.*

Satellite images (below) show the nightside half of the auroral oval during a substorm, representing an intense westward traveling surge. When a westward traveling surge moves along the auroral oval, an auroral curtain often becomes folded. From a satellite altitude, the folded portion looks like a loop (below, bottom).

An intense westward traveling surge at Fort Yukon is illustrated by a series of all-sky camera photographs in the image at right. The image at lower right illustrates a view of the same event from a satellite (the field of view of the Fort Yukon all-sky camera is shown by the circles).

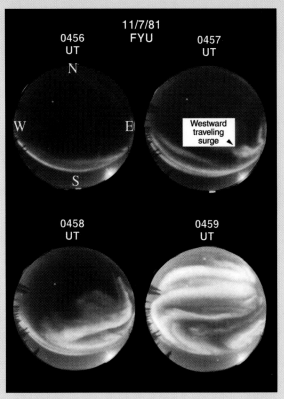

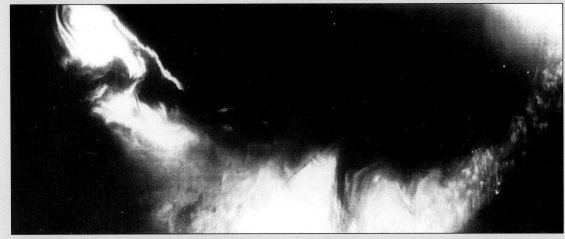

▲▼ COURTESY OF DEFENSE METEOROLOGICAL SATELLITE PROGRAM

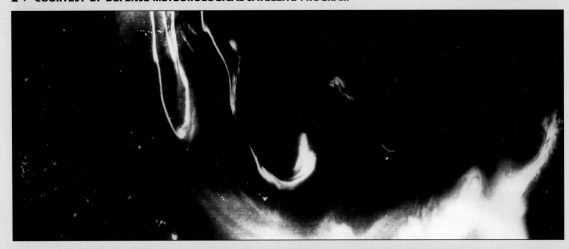

▲▼ FROM S.-I. AKASOFU

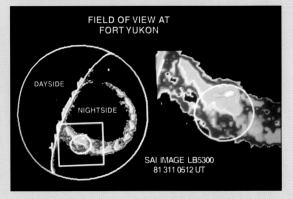

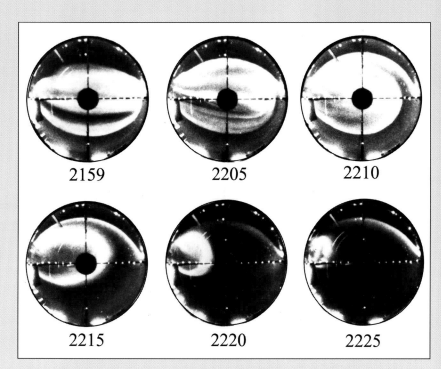

A series of all-sky photographs shows a looplike structure as it drifted toward the western horizon (above). Photographs at right, taken with a fish-eye lens, show the westward-drifting loop.

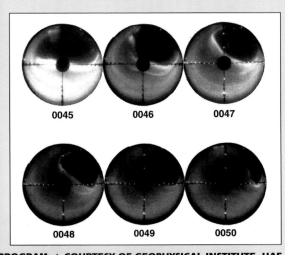

0045 0046 0047

0048 0049 0050

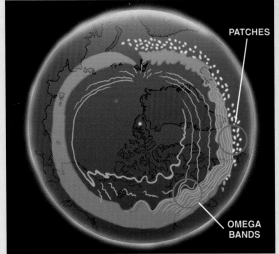

PATCHES

OMEGA
BANDS

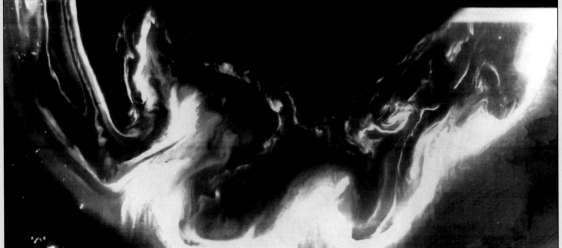

FACING PAGE: *Auroral displays in the morning sky are different from those in the evening sky. An auroral curtain shows an inverted omega-shaped fold, called the omega band (upper left). A series of all-sky photographs (upper right) shows an eastward-shifting omega band (the time is in Alaska Standard Time). When you watch auroral displays in the morning sky, an auroral curtain becomes bright near the western horizon first. Then a series of omega bands passes by you and shifts toward the eastern horizon. At lower left, an illustration shows where omega bands and patches are located. At lower right, a satellite image shows auroral activity in the nightside half of the auroral oval. You can see a loop in the evening sky, while omega bands and torches (see page 86) are seen in the morning sky.*

◀ **JACK FINCH**
▼ **COURTESY OF GEOPHYSICAL INSTITUTE, UAF**

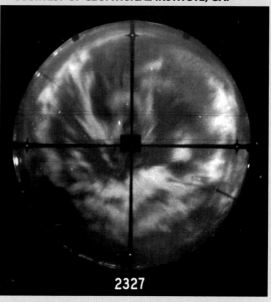

2327

LEFT AND ABOVE: *In the morning, patchy forms are scattered all over the sky (left). The patches drift slowly eastward (toward the morning twilight sky). Many of the patches pulsate; their brightness changes periodically, with a period of several to more than 10 seconds. Above is an all-sky photograph of the patchy aurora.*

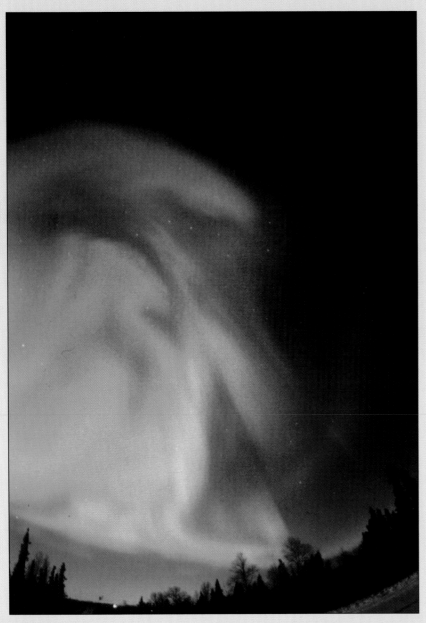

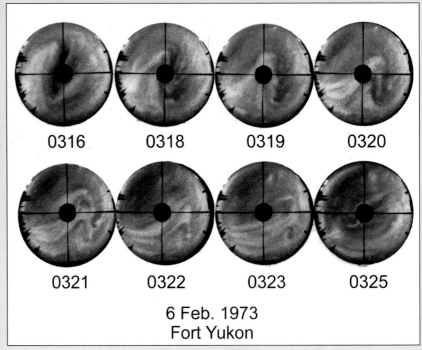

0316   0318   0319   0320

0321   0322   0323   0325

6 Feb. 1973
Fort Yukon

In the morning sky, the aurora often shows a form called the torch (left); it develops from a diffuse glow located a little southward of the curtainlike aurora (see the lower right image on page 84). A series of all-sky photographs, above, shows the torch structure as it drifted eastward (the time is in Alaska Standard Time).

We have come a long way from the days of Sophus Tromholt, who noted, "Who will venture to answer to the question 'What is the aurora?'"

It has now become clear that auroral lights appear when atoms and molecules in the upper atmosphere are hit by high-speed electrons coming from above. Indeed, this chapter will show that the aurora is a gigantic electrical-discharge phenomenon surrounding Earth. How is this electrical discharge powered? What kind of processes in the sky can act like a generator, supplying the electricity to this most spectacular natural phenomenon? The electrical power associated with the auroral discharge is enormous, about 1,000,000 megawatts or more. This power is greater than the total U.S. electric power consumption.

The ultimate source of this power is the

Observed from almost directly beneath this display in Turnagain Pass, about halfway between Anchorage and Seward, the aurora's rays appear to converge. (Daryl Pederson)

The solar corona, the outermost solar atmosphere, can be seen in this photo taken during a total eclipse of the Sun. (Courtesy of S. Numazawa and K. Kozuka; from S.-I. Akasofu)

The story of the search for this particular Earth-Sun relationship began at 11:20 A.M. on September 1, 1859, when Richard C. Carrington, an English astronomer and solar physicist, was sketching sunspots. To his surprise, a very bright spot appeared among the spots and he made the first notation of what we now call a solar flare, an intense explosion on the Sun. Carrington "hastily ran to call someone to witness the exhibition." About one day later, a large part of Europe was covered by active auroras. In fact, some records indicate that the aurora was seen as far south as Honolulu on that day. Simultaneously, major geomagnetic disturbances also occurred. Carrington suggested that there was a possible connection between the solar explosion he had witnessed and the great auroral display, but he also warned against jumping to conclusions by remarking, "One swallow does not make a summer."

Many physicists and astronomers gave more heed to his warning than his suggested connection. Even in 1895, William Thomson, Lord Kelvin, one of the most famous British physicists of his day, commented that Carrington's finding was simply coincidental and called the problem a "'fifty years' outstanding difficulty."

By the end of the nineteenth century,

Sun. The solution to the problem of the relationship of the aurora to the Sun, however, did not come easily. It is based on efforts made by many scientists over the last 100 years. Even as recently as the beginning of the twentieth century it was not at all obvious that the aurora is related to the Sun.

many scientists, including Alexander von Humboldt (who coined the term "magnetic storm"), Anders Celsius (who designed the centigrade scale for thermometers), Kristian Birkeland, and others found a close relationship between auroral displays and geomagnetic disturbances. In 1905, after an exhaustive study of geomagnetic disturbances, another British solar physicist Edward W. Maunder stated, "The origin of our magnetic disturbances lies in the Sun."

However, his colleague, the mathematician and physicist Sir Arthur Schuster, immediately criticized him, saying, "The mystery is left more mysterious than ever."

Many more years had to pass before

The magnetosphere, an electric generator, looks like a cylinder with a blunt nose. This is because solar wind particles blow around the magnetosphere, forming a comet-shaped cavity, with Earth located near the head of the cavity. The cavity extends away from the Sun at least the length of 1,000 Earth radii, or 4,000,000 miles. This portion is called the magnetotail. This illustration shows the magnetotail cut open to reveal its internal structure, together with the boundary, called the magnetopause, and the motions of solar wind particles there. Solar wind particles consist of protons (+) and electrons (−). As they blow along the magnetopause, they flow across the magnetic field there, generating electrical power. As a result, the morningside of the boundary is positively charged, while the eveningside is negatively charged. (S.-I. Akasofu)

the majority of scientists became convinced of this particular solar-terrestrial relationship, despite the fact that it had been well established in the latter part of the nineteenth century that the occurrence of the aurora has a similar cycle to that of the sunspot cycle. In 1884 Sophus Tromholt quipped, "How can this connection between what may be called a terrestrial phenomenon and solar disturbances be explained? Well, that is one of those riddles which the scientist of today leaves to that of the future to solve."

It turns out that the answer to this riddle is quite complicated, but a solution may now be coming into sight.

The Solar Wind

The Sun is a ball of hydrogen gas. In its central core, both temperature and pressure are high enough to produce a nuclear fusion reaction. The energy generated by the reaction flows out from the core to the photosphere, the visible disk of the Sun. The photosphere has a temperature of

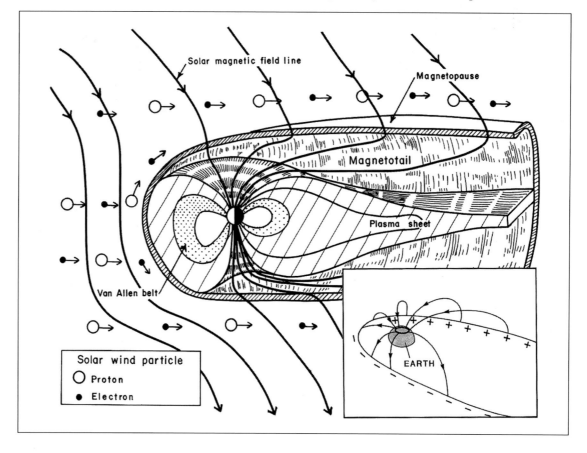

about 6,000 degrees Kelvin (0°C = 273°K). The photosphere is the lowest part of the solar atmosphere. Every living creature on Earth relies on the energy emitted by the photosphere. For the aurora, the Sun generates another source of energy. Above the photosphere, there are two layers of atmosphere, the chromosphere and the corona. Both can be seen during a solar eclipse, when the bright photosphere is covered by the disk of the moon.

The corona, the outermost atmosphere of the Sun, has a temperature of about 1,000,000 degrees Kelvin. The cause of the heating process of the corona is not known. Because of this high temperature, the corona cannot remain near the Sun and expands with a speed of a few hundred miles per second. This coronal flow is called the solar wind. This hot wind blows all the way to the edge of the solar system, passing by Earth, the outer planets, Jupiter and Saturn, Uranus, Neptune, and Pluto. Since the corona is so hot, its main constituents, hydrogen atoms, are ionized, namely each hydrogen atom is split into a proton and an electron. Therefore, the solar wind is mostly a flow of both protons and electrons. Such an ionized gas is called plasma.

The Magnetosphere

When the solar wind, a stream of protons and electrons, blows by Earth, it forms a cometlike cavity around the planet. However, because Earth is a magnet, the solar wind cannot penetrate deep toward

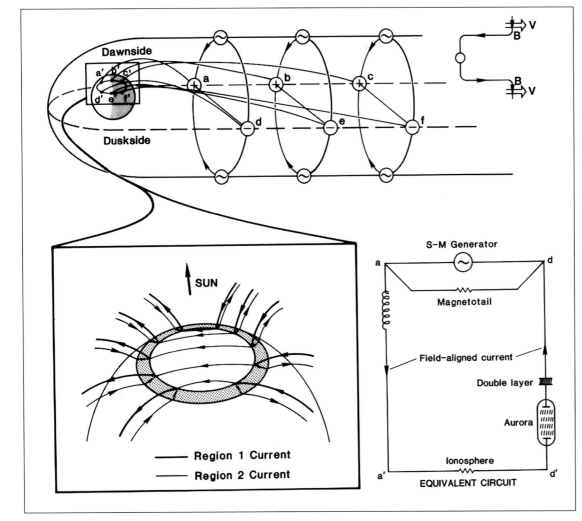

This illustration shows the auroral electrical circuit. The morningside of the magnetopause becomes the positive "terminal" of the solar wind-magnetosphere generator and the eveningside becomes the negative "terminal." The electric current flows from the positive terminal to Earth's polar upper atmosphere in the morningside and back from Earth to the negative terminal in the eveningside, from a to a¹, a¹ to d¹, and d¹ to d; also from b to b¹, b¹ to e¹, and e¹ to e, and so on. An equivalent current diagnosis is also shown. This portion is called the Region 1 Current; there is also a secondary current called the Region 2 Current. (S.-I. Akasofu)

Earth is not the only planet with auroras. If present understanding of the auroral processes is correct, the magnetized planets (such as Jupiter, Saturn, and Uranus) should have the aurora, while nonmagnetized planets (such as Venus and Mars) should not have the phenomenon. (By NASA's Hubble Space Telescope, courtesy of A. Bhardwajal and G.R. Gladstone; from S.-I. Akasofu)

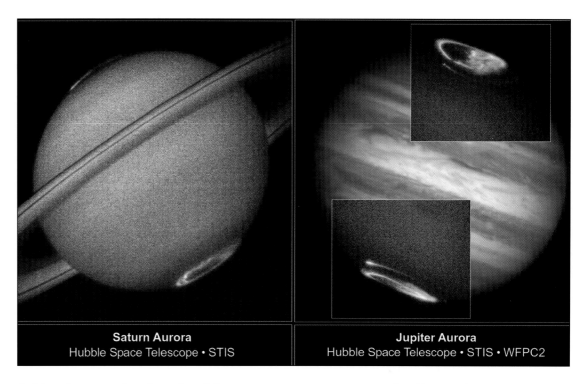

Saturn Aurora
Hubble Space Telescope • STIS

Jupiter Aurora
Hubble Space Telescope • STIS • WFPC2

Earth and tends to flow around it at a distance of about 10 Earth radii (40,000 miles). This cavity created by the diverted flow, with a long tail of about 1,000 Earth radii (4,000,000 miles), is called the magnetosphere.

The solar wind carries out magnetic fields from the photosphere. The magnetic fields in the solar wind interact with Earth's magnetic field. In fact, the magnetic field lines of the solar wind and Earth's magnetic field lines connect across the boundary of the magnetosphere.

The Solar Wind-Magnetosphere Generator

A generator is a simple machine that consists of a coil and a magnet, converting the energy of the rotating coil into electrical energy. To rotate the coil in the magnetic field, hydropower or steam power is used. In turn, to generate steam power, coal and oil have been used. As a result, carbon dioxide is released into the atmosphere, contributing to global warming.

What's important in a generator is a conductor (coil) moving in a magnetic field. What can play the role of the conductor and the magnet in generating power for the auroral electrical discharge?

As the solar wind blows around the magnetosphere, both protons and electrons must move along the boundary of the magnetosphere across the magnetic field lines that result from the connection of the solar wind magnetic field and Earth's magnetic field. Thus, the solar wind blowing along the boundary of the magnetosphere becomes a generator and produces the power needed for the auroral discharge. Auroral scientists have succeeded in estimating the power to be at least 1,000,000 to 10,000,000 megawatts, the voltage being about 30 to 100 kilovolts, and the electric currents tens of millions of amperes. The largest power generator in the world can generate about 1,000 megawatts; our auroral generator can produce 1,000 times more power than that.

Auroral Discharge Circuit

The positive terminal of the solar wind-magnetosphere generator is distributed along the morningside of the equatorial plane of the boundary of the magnetosphere, while the negative terminal is distributed along the eveningside.

Part of the electric current created by the generator flows directly between the two terminals across the tail. The auroral

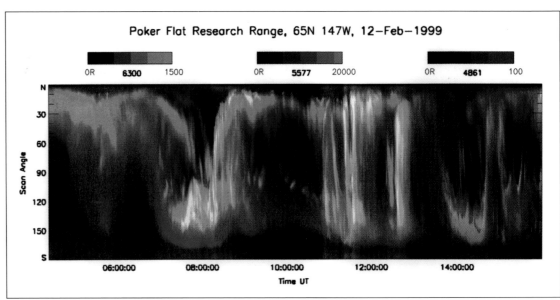

Poker Flat Research Range, 65N 147W, 12-Feb-1999

0R 6300 1500
0R 5577 20000
0R 4861 100

LEFT: *Experiments using rockets launched from Poker Flat Research Range near Fairbanks have yielded important information for auroral scientists. Since the Range was established 33 years ago, more than 1,500 meteorological rockets and 250 major high-altitude sounding rocket experiments have been launched to conduct atmospheric research on the aurora, the ozone layer, solar protons and electrons, magnetic fields, and infrared, ultraviolet and x-rays. Researchers of the Geophysical Institute, University of Alaska are now testing the effects of the aurora on GPS and other radio systems. (Courtesy of Geophysical Institute, University of Alaska; from S.-I. Akasofu)*

ABOVE: *Auroral data from February 12, 1999, includes the auroral emissions (the red atomic oxygen line, the green atomic oxygen emission, and the blue-violet ionized nitrogen molecular emissions) from the northern horizon (0°), the zenith (90°), and the southern horizon (180°) as a function of Universal Time. (Courtesy of Geophysical Institute, University of Alaska; from S.-I. Akasofu)*

discharge current flows from the positive terminal toward the polar region of Earth along the magnetic line of force and flows back to the negative terminal after flowing across the polar ionosphere.

In the magnetosphere, which is filled with rarefied ionized gases (only about 100 particles in one cubic inch, compared with billions and billions in Earth's atmosphere near the ground), the current-carrying electrons are constrained to move only along the magnetic field lines. Thus, the magnetic field lines act like invisible wires, connecting the terminals and the upper atmosphere where the discharge current can flow. This layer of the atmosphere is called the ionosphere, which is located at an altitude of 60 to 70 miles.

As the positive terminal is a line rather than a point, a large number of "wires" from the positive terminal land along two narrow belts in the morningside of the polar ionosphere, one in the Northern Hemisphere and the other in the Southern

Hemisphere. These belts coincide with the morning half of the auroral oval in the Arctic and the Antarctic, respectively. The "wires" from the evening half of the auroral

Researchers monitor auroral conditions from inside the Neil Davis Science Operations Center at Poker Flat Research Range, the largest land-based rocket range in the world. Auroral and magnetic conditions at the Range, Fort Yukon, Kaktovik, and other places are displayed. (Courtesy of Geophysical Institute, University of Alaska; from S.-I. Akasofu)

oval in both hemispheres are connected to the negative terminal. This is why the aurora appears along the auroral oval, not everywhere in the polar sky. The amount of this discharge current in the ionosphere is several million amperes.

The discharge currents are carried mainly by electrons. In this process, the current-carrying electrons collide with upper-atmospheric atoms and molecules. The collisions provide energy for the atoms and molecules to emit lights that we recognize as the aurora. By definition, the electrons flow in the direction opposite to

that of the currents. Thus, the auroral electrons flow from the negative terminal to the evening half of the auroral oval. This explains the aurora in the evening half of the oval.

In this process of carrying the auroral discharge current, there are two mysteries that have not been solved yet. The first is that the current-carrying electrons are accelerated downward, toward Earth. In each collision, an electron loses tens of electron volts of energy. For the electrons to penetrate down to an altitude of 60 miles, the bottom edge of the auroral curtain, the electron has to collide hundreds of times with upper atmosphere atoms and molecules. Therefore, the electrons have to have several kilo electron volts to reach an altitude of 60 miles. The electrons in the magnetosphere have only a few electron volts of energy, so that they have to be accelerated thousands of times. In spite of many years of intensive research with rockets and satellites, auroral scientists

PLANET	MAGNETIC FIELD	ATMO-SPHERE	AURORA EXPECTED	AURORA OBSERVED
Mercury	YES	NO	NO	NO
Venus	NO	YES	NO	NO
Earth	YES	YES	YES	YES
The Moon	NO	NO	NO	NO
Mars	NO	YES	NO	NO
Jupiter	YES	YES	YES	YES
Saturn	YES	YES	YES	YES
Uranus	YES	YES	YES	YES
Neptune	YES	?	?	?
Pluto	?	?	?	?

have not conclusively determined the cause for this acceleration process. The second mystery is why this discharge current forms a thin sheet or sheets. That is to say, the current-carrying electrons flow in a very thin sheet or sheets. This is why the aurora looks like a curtain of lights in the sky.

In the ionosphere, ionospheric electrons take over the task of carrying the currents from the evening half of the oval to the morning half of the oval. From the morning half of the auroral oval, ionospheric electrons carry the currents to the positive terminal; the ionospheric electrons are not energetic enough to produce auroral lights. Actually, the discharge current circuit is much more complicated than what is described here, and there is another discharge circuit that brings electrons in the magnetosphere to the morning half of the oval, causing the aurora in the morningside of the oval as well.

The above description of the cause of the aurora is quite different from what is presented in many books. It is generally believed that the aurora is caused by solar particles that are attracted by Earth's magnetic field and impact on the polar upper atmosphere, producing the aurora. Such an idea cannot explain a simple fact that the aurora appears along the auroral oval. The story is much more complicated. As we learned earlier, the aurora is associated with an electrical discharge, which is powered by the solar wind-magnetosphere generator. The important point is that the aurora is not caused by a direct impact of solar particles from the Sun on the polar upper atmosphere.

The auroral discharge currents generate complicated magnetic fields that can be observed by an instrument called a magnetometer, more than 50 of which are located in northern high latitudes and others that are carried by satellites. In fact, auroral scientists have learned about the auroral discharge circuit by analyzing the magnetic fields produced by these currents.

It is now possible to test the present understanding of the cause of the aurora in an exciting way. If the present understanding is correct, planets with a magnetic field and atmosphere should have the auroral oval since the solar wind blows beyond the distance of all

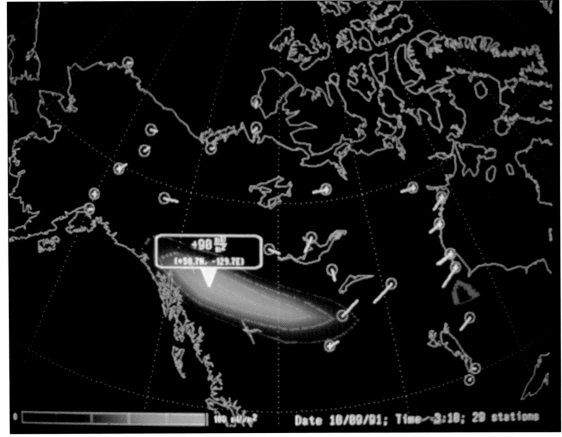

Auroral data displayed on a computer monitor shows the direction and magnitude of magnetic fields produced by auroral activity at a number of magnetic observatories (indicated by a circle) in North America. The red area shows the location where the upper atmosphere is heated by the auroral discharge current. (Courtesy of Geophysical Institute, University of Alaska; from S.-I. Akasofu)

The intense electrical discharge currents that cause the aurora can also cause various problems for man-made objects on the ground and in space. The auroral currents induce currents in a powerline system, sometimes causing a blackout. Oil/gas pipelines such as the trans-Alaska oil pipeline are protected to avoid corrosion from the induced current. Satellites can also be disturbed as they go through auroral currents. (Greg Syverson)

SCIENTIFIC INSTRUMENTS	
Magnetometers	Measure changing magnetic fields to infer the discharge currents and their circuits.
All-Sky Cameras	Photograph the whole sky to record auroral displays.
Photometers	Measure the intensity of auroral lights. Various filters are attached to study variations of the intensity of a particular wavelength of light.
Radars	Various types of radars are used to measure electric field electron density.
Rockets and Satellites	Carry magnetometers, photometers, imaging devices, particle detectors, and devices to measure electric field, etc.

the planets in the solar system.

So far, the test is satisfactory. It is interesting to note that the aurora on Jupiter and Saturn is pink, because their atmospheres consist mainly of hydrogen. Earth has a greenish-white aurora because the main emission comes from atomic oxygen. Oxygen atoms are released as a by-product of photosynthesis of plants. Therefore, if planets of stars other than the Sun have the oxygen emission, there is a good chance that there is at least plant life there. In the future, auroral research could develop into search for life on planets or stars.

Scientific Instruments

Auroral scientists use a variety of instruments in studying the aurora. There are a number of auroral observatories in the polar regions, in both the Arctic and the Antarctic, which are equipped with a great variety of these. The Poker Flat Research Range of the Geophysical Institute, University of Alaska is one such observatory. The Range is equipped with several powerful rocket launchers as well. There are many scientists at different universities who construct instruments carried by rockets and satellites. ■

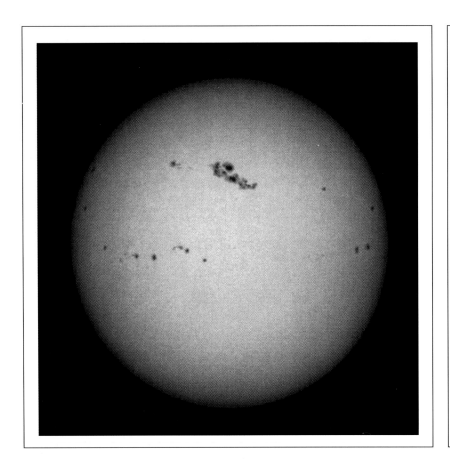 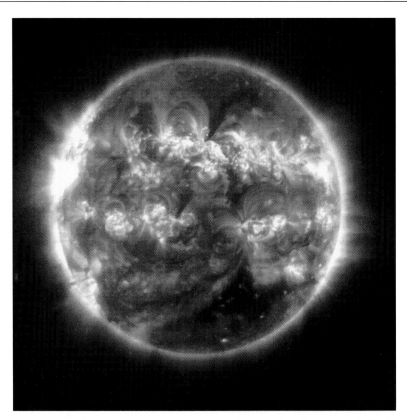

The ultimate source of energy for the aurora is the Sun, or more specifically, the motion energy carried by solar wind protons. An electric generator converts motion energy into electric energy. The solar wind-magnetosphere generator coverts the motion energy of solar wind protons into electrical energy for auroral phenomena.

A hydropower generator can produce more power when water flow is high. Similarly, when the solar wind speed is high, the solar wind-magnetosphere generates more electric power. A change in the brightness of the aurora reflects in part changes in speed of the solar wind. Auroral displays are a reflection of changes on the Sun.

Actually, the situation is a little more complicated. As mentioned in the previous section, the auroral electric power is generated as the solar wind blows around the comet-shaped boundary of the magnetosphere across the combined magnetic field

FACING PAGE, LEFT: *A large sunspot group can be seen at the central meridian of the solar disc in this image recorded on March 28, 2001. (Courtesy of Big Bear Solar Observatory; from S.-I. Akasofu)*

FACING PAGE, RIGHT: *An x-ray image obtained by the SOHO satellite shows sunspot activity. (Courtesy of NASA; from S.-I. Akasofu)*

RIGHT: *The Alyeska Prince Hotel, 40 miles south of Anchorage, lights up the night under an active auroral curtain. (Daryl Pederson)*

of the solar wind magnetic field and Earth's magnetic field. Because the intrinsic magnetic field of Earth is stable, changes of the combined field on the boundary of the magnetosphere depend on changes of the solar wind magnetic field. The orientation of Earth's magnetic field is such that the

number of the combined magnetic field lines is greater when the magnetic field of the solar wind is oriented southward rather than northward. A stronger combined field allows a higher power for the generator. Therefore, the power of the solar wind-magnetosphere generator depends on both the solar wind speed and the solar wind magnetic field magnitude and orientation. Auroral

scientists determined the formula for the power:

Power (megawatt) = 20x [solar wind speed (km/sec)] x [solar wind magnetic field magnitude (nano Tesla)]^2 x $\sin^4(\theta/2)$

Here q is the polar angle of the solar wind magnetic field ($\theta = 0°$ for a north-ward field and $\theta = 180°$ for a southward

BELOW: *Scientists observed a large coronal hole (dark area) on the Sun in August 1974. (Courtesy of NASA Skylab Project, the Solar Physics Group, American Science and Engineering Inc.; from S.-I. Akasofu)*

RIGHT: *Sunspot activity follows an 11-year cycle (top); auroral activity (bottom) follows the sunspot cycle but tends to lag behind by a few years because coronal holes, which produce solar wind, tend to develop a few years after the year of sunspot maximum. (S.-I. Akasofu)*

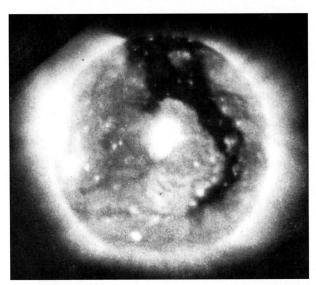

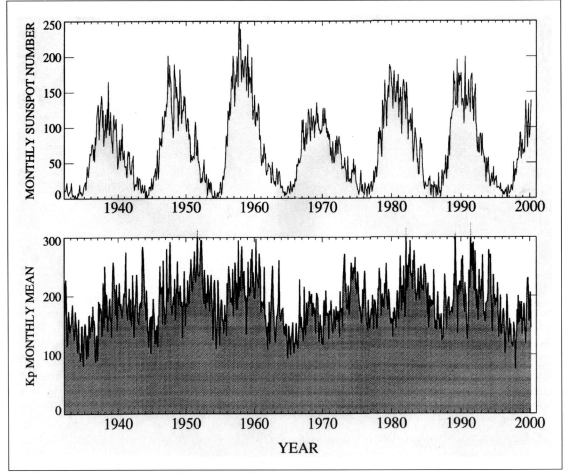

field). In the above three quantities, the polar angle θ is most variable in the solar wind. In fact, when q varies, say, from 30° to 150° for three hours or so, a typical auroral substorm occurs. Auroral scientists call this change of the solar wind magnetic field angle "a southward turning."

When we observe the Sun visually, all we can see is a bright disk and sometimes a few to several dark spots called sunspots. However, when we observe the solar corona by an x-ray image aboard a satellite, we see an entirely different face of the Sun. There are a great variety of fine, bright structures, dark spaces (called coronal holes), and many small bright spots. They change almost continuously. Further, the sun rotates every 25 days almost like a rigid ball. Therefore, we observe different faces of the Sun throughout the 25 days. This is why the solar wind speed and magnetic field vary almost continuously at Earth (or elsewhere).

In this complex situation, both solar physicists and auroral scientists found that the solar wind varies in at least two ways. When the Sun was imaged by an x-ray detector, it was found that a large coronal hole produces a strong solar wind. This is one of the mysteries of the Sun; why does it produce a strong solar wind from a dark (thus cooler) region of the corona?

The Sun rotates every 25 days and the solar wind from the coronal hole rotates with the Sun, just like a beacon at an airport. Therefore, the solar wind collides with the magnetosphere every 25 days; during one solar rotation, Earth is also rotating around the Sun in the same

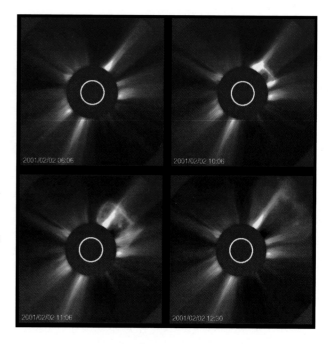

A coronal mass ejection (CME) billows from the Sun on January 28, 2001. The white circle is an outline of the Sun behind an "artificial moon," which is used as a kind of viewing shield against the intense light. (Courtesy of NASA; from S.-I. Akasofu)

direction that the Sun is turning. Thus Earth advances in the direction of the rotation of the solar wind stream by an amount of two days during the solar rotation period of 25 days. Therefore, the solar wind collides with the magnetosphere every 27 days. The stream from a coronal hole is wide and it takes one week to 10 days to pass by Earth. Thus, auroral activity lasts for one week to 10 days during the period when Earth is immersed in the strong, turbulent wind from a coronal hole. The Sun has a cycle of activity called the solar cycle, which has a period of about 11 years. The number of sunspots waxes and wanes in 11 years, and coronal holes tend to develop a few years after the year of sunspot maximum.

The Sun also displays intense transient activities when a large eruption occurs around a large sunspot group. The eruption is associated with a sudden brightening of the chromosphere, the atmospheric layer below the corona. This phenomenon is called a solar flare. The corona above the flare region exhibits complicated changes. Some of the large eruption causes a part of the corona to be blown away, a phenomenon called the coronal mass ejection. The cause of this phenomenon is not known.

It is expected that the CME advances away from the Sun with a speed of 500 miles per second or greater into a slower solar wind ahead of it. This action creates a shock wave in the solar wind just like a bullet moving in the air. As a result, both CME and the shock wave advance together into space between the Sun and Earth and beyond. It takes, on average, two days for the CME shock to reach Earth's orbit.

As the shock wave reaches the front of the magnetosphere, it compresses the magnetosphere. The generated magnetic pulse reaches Earth's surface in about one minute. As the shock wave is associated with a slightly higher solar wind speed than a slow wind, the power of the auroral generator increases a little after the passage of the shock wave.

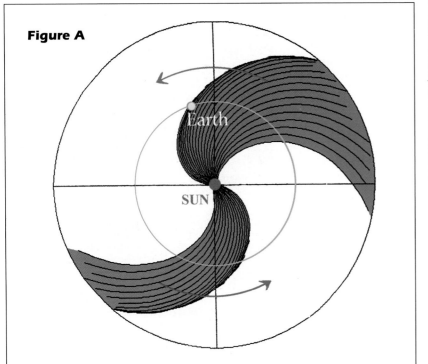

Figure A

Earth

SUN

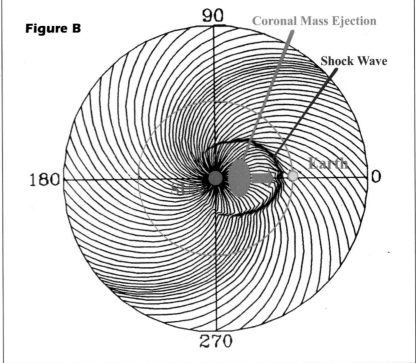

Figure B

90

Coronal Mass Ejection

Shock Wave

180

Earth

0

270

The Sun has two modes of blowing its uppermost atmosphere, the corona. The first one (Figure A) is a fairly steady flow (blue portion) from coronal holes, usually in two streams. The flow rotates with the Sun like a beacon at the tower of an airport with a period of about 25 days. Because Earth rotates in the same direction, the same flow engulfs it every 27 days. The aurora is active during the seven to 10 days the flow needs to overtake Earth in each solar rotation. The second mode is like a gusty wind (Figure B) when a CME blows out from an active region around sunspots. An intense flow lasts only about 24 hours, so that the associated auroral activity, though intense, lasts only that long. (S.-I. Akasofu)

Subsequent changes in the power of the generator and of auroral displays depend on how the solar wind speed and its magnetic field magnitude and orientation vary within the CME. Even if the solar wind speed and the magnetic field magnitude substantially increase, the generated power remains low if the polar angle θ is small. In fact, if the field is oriented northward (sin θ = 0°), our generator cannot generate any power.

There is one more complication. The generator power increases greatly when the shock wave and CME "hit" the magnetosphere in the direction of its advance, a "direct hit." This happens when a solar flare/CME occurs near the center of the solar disk. If a solar flare/CME occurs near the edge of the solar disk, the CME advances 90° away from the Sun-Earth direction, so that only the skirt of the shock wave may reach the magnetosphere. Therefore, there are many reasons why an intense solar flare does not necessarily produce a great auroral display.

In predicting auroral activity, we have to be able to predict in advance the solar wind speed, magnetic field magnitude, orientation, and their variations in time. Auroral scientists, together with solar physicists, are working hard in what we call space weather prediction.

Great Auroral Displays and Geomagnetic Storms

When an intense CME occurs near the center of the solar disk and its magnetic field is strong and oriented southward, the power of the solar wind-magnetosphere generator may exceed 10,000,000 megawatts. As a result, a great auroral display occurs. It is difficult to predict such events, since the cause of solar flares is not well known. Simultaneously, the magnetic field produced by the auroral discharge current causes a great geomagnetic storm. Both an auroral display and a geomagnetic storm are different aspects of the same phenomenon, a great increase in the power of the solar wind-magnetosphere generator. In such a situation, the upper atmosphere is

greatly heated. When oxygen atoms collide with heated electrons, the atoms emit a dark red light. This emission tends to occur at an altitude of 150 miles and above. Further, the auroral oval shifts toward the equator. Therefore, it can be seen from a great distance in lower latitudes. It is this red glow that caused a great fear among people in ancient and medieval times. This red aurora has a tendency to occur during sunspot maximum years, because an intense CME tends to occur around a large sunspot group.

For example, during the great geomagnetic storm of February 11, 1958, the oval shifted at times even south of the Lower 48 United States-Canada border, so that the aurora temporarily disappeared from the polar sky. The Alaska sky was

almost completely deserted at about 10:00 UT on February 11, 1958. However, that situation did not last long. About half an hour later, auroras spread over a wide belt covering much of the northern United States and Canada.

BELOW, LEFT: *This time sequence illustration shows auroral activity on July 15, 2000. (Courtesy of NASA/IMAGE FUV Team; from S.-I. Akasofu)*

BELOW: *During a solar flare, huge chunks of hydrogen gas are blown from the Sun. This causes a CME, in which part of the corona is also blown outward, creating a giant shock wave that moves through space. (Courtesy of NASA; from S.-I. Akasofu)*

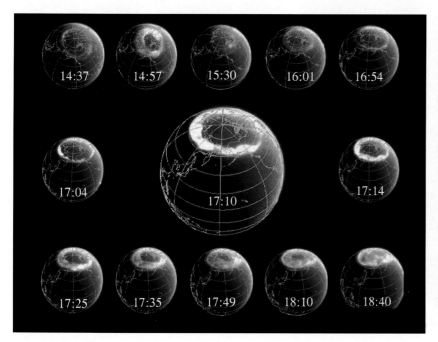

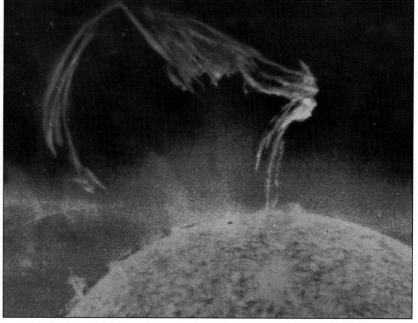

The top of auroral curtains during such a great display tends to be as high as 600 miles and is a rich, dark red color, emitted by atomic oxygen; atomic oxygen emits the dark red line when it collides

Active auroras tend to interfere with short wave (HF) radio communications because auroral electrons produce ionization at about 55 miles in altitude, a little below the height of the aurora. The ionization there tends to absorb radio waves rather than reflecting them. (Jan Curtis; from S.-I. Akasofu)

with heated (low energy) electrons, while it emits the green line when it collides with discharge (high energy) electrons. Because of its high altitude, the light can be seen from a great distance, from much farther south than usual. On February 11, 1958, people in Mexico saw the red aurora. Similar great auroral displays were seen on September 1, 1859, in Honolulu; February 4, 1872, in Bombay (Mumbai); September 25, 1909, in Singapore; May 13, 1921, in Samoa; and in Mexico on September 13 and 23, 1957.

Viewers wishing to see the aurora at

high latitude locations, such as Fairbanks (Alaska), Tromso (Norway), Yellowknife (Canada), and others, have a much better chance a few years after the year of a sunspot maximum; namely 2004 to 2007, during the present sunspot Cycle 23; and 2015 to 2018, during the next sunspot Cycle 24. This is because coronal holes tend to develop a few years after the year of the sunspot maximum. Earth and its magnetosphere are immersed in the coronal hole stream for one week to 10 days during which auroral activity is high. Also, it is possible to predict auroral activity at least 27 days ahead. On the other hand, it is difficult to predict solar flares and CMEs, and it is not certain which flares will produce a great display. Further, CMEs pass by Earth in less than one day and intense displays last only for about 24 hours.

Auroral scientists are now working with solar physicists to provide reliable forecasts to power companies and communications and defense facilities. There is a worldwide solar monitoring system that watches the Sun continuously with various optical, radio, and satellite-borne instruments. The forecast is issued from the Solar Forecasting Center, Space Environmental Laboratory, the National Oceanographic and Atmospheric Administration, in Boulder, Colorado. Because of these wide-ranging effects of the phenomenon, there are a number of research institutes in the world that dedicate major effort to studying the aurora. The Geophysical Institute at the University of Alaska is one of them. ∎

Photographing the Aurora

To capture the northern lights on film, you will need the following camera equipment: a sturdy tripod, a locking-type cable release (some 35mm cameras have both *time* and *bulb* settings, but most have *bulb* only, which calls for use of the locking-type cable release), and a camera with an f/3.5 lens (or faster). A few photographers get good results shooting the aurora with high-end digital cameras, but single-lens reflex (SLR) cameras tend to offer better detail and resolution than the average digital camera and are therefore still widely used for this type of sensitive photography.

It is best to photograph the lights on a night when they are not moving too rapidly. As a general rule, photos improve if you manage to include recognizable subjects in the foreground — trees and lighted cabins are favorites of many photographers. Set your camera up at

Using a tripod to photograph the aurora is necessary for the long exposures required to record these ethereal northern lights. (Cary Anderson)

LEFT: *Aurora photographers can find information online about current auroral activity and predictions for viewing possibilities in their area. (Daryl Pederson)*

BELOW: *Since the aurora is only visible during nights when the sky is clear, the cold is a factor for photographers when dressing for a shooting session and setting up camera equipment outside. (Cary Anderson)*

SUGGESTED EXPOSURE TIMES		
f-stop	ASA 200	ASA 400
F1.2	3 seconds	2 seconds
F1.4	5 seconds	3 seconds
F1.8	7 seconds	4 seconds
F2	20 seconds	10 seconds
F2.8	40 seconds	20 seconds
F3.5	60 seconds	30 seconds

least 75 feet from the foreground objects to make sure that both the foreground and aurora are in sharp focus.

Normal and wide-angle lenses are best. Try to keep your exposures under a minute — 10 to 30 seconds generally works best. The lens openings and exposure times shown in the chart above are only a starting point, since the amount of light generated by the aurora is inconsistent. (It's best to bracket exposures.)

Ektachrome and Fujichrome 200 and 400 color slide films can be push-processed in the home darkroom or by some custom-color labs, allowing use of higher ASA ratings (800, 1200, or even 1600 on the 400 ASA film, for example). Consult your local camera store for details. Once processed, prints or slides can be scanned; the resulting digital files can then be manipulated on a computer using image-editing software.

A few notes of caution: Remember to protect the camera from low temperatures until you are ready to make your exposures.

Some newer cameras have electrically controlled shutters that will not function properly at low temperatures. Also, wind the film slowly to reduce the possibility of static electricity, which can lead to streaks on the film. Grounding the camera when rewinding can help prevent the static electricity problem. (To ground the camera, hold it against a water pipe, drain pipe, metal fence post, or other grounded object.)

Follow these basic rules, experiment with exposures, and you should obtain good results. ■

Photographer Patrick J. Endres of Fairbanks took this photo (also used on our cover) near the head of Canwell Glacier, east of the Richardson Highway in the Alaska Range, while camping in April 2000. A full moon lit the surrounding peaks and the nighttime temperature hovered at zero degrees Fahrenheit. Around 1 A.M., Patrick used a Canon EOS 3 with a 24mm f1.4 lens to capture auroral displays on Fuji 100 Provia slide film. This exposure lasted about 10 seconds with the lens set at f1.4. The lab then push-processed the roll. No digital enhancements or edits were done to the image. (Patrick J. Endres)

Glossary

all-sky camera: a camera that is capable of photographing the entire sky in a single frame.

atomic oxygen: oxygen molecules (O_2 in the lower atmosphere) that are separated into their two oxygen atoms (O) at the altitude of the aurora. Atomic oxygen emits a whitish-green light and a "bloody red color," depending on conditions.

aurora: visible light emitted by gases when struck by fast-moving electrons and protons in the high atmosphere (above 60 miles altitude) over Earth's polar regions.

auroral curtain: the basic auroral form, appearing as a curtain of light.

auroral oval: a narrow band that encircles the geomagnetic poles along which auroral curtains lie. The aurora in the northern oval is called the aurora borealis, while the aurora in the southern oval is called the aurora australis.

auroral spectroscopy: a field of science that studies light emitted by the aurora.

auroral substorm: intense auroral activity all along the auroral oval, lasting for a few hours.

auroral zone: the belt around the polar regions where the aurora can be seen the maximum number of nights per year, about 200-250 in the northern polar region.

chromosphere: the layer of the solar atmosphere between the photosphere (visible disk of the Sun) and the corona (outermost solar atmosphere).

corona, auroral: an auroral form seen when an active auroral curtain is directly overhead; the light rays appear to converge near the zenith.

corona, solar: the outermost layer of solar atmosphere; its temperature is a few million degrees Kelvin.

coronal holes: dark spaces observed when viewing the solar corona via a satellite x-ray image. Large coronal holes produce strong solar winds.

coronal mass ejection (CME): a phenomenon whereby a part of the solar corona is blown away during an intense solar activity.

diffuse aurora: a faint belt of light with a uniform brightness that appears just southward of the auroral oval; looks similar to the Milky Way.

excitation: a process in physics whereby an atom's internal state changes from normal to a higher level of energy due to energetic electrons colliding with it; in the case of the aurora, the resulting energy is emitted as light.

geomagnetic pole: a hypothetical point on Earth's surface (representing a line going through a hypothetical bar magnet at the planet's center and extending to Earth's surface; the penetrating point of the line is called the geomagnetic pole). In the Northern Hemisphere, the geomagnetic pole is located at Ellesmere Island in Canada; in the Southern Hemisphere it is at Vostok in Antarctica.

green line: the most common light of the aurora, a whitish-green color, emitted by atomic oxygen, the wavelength being 557.7 nm.

ground state: the normal condition of an atom or molecule.

homogeneous arc: a quiet aurora characterized by a uniform brightness in a horizontal direction.

inverted omega band: an auroral curtain that tends to develop a particular shape of fold that projects onto the ground like the inverted Greek character Ω in the morning sector.

ionosphere: the region in Earth's atmosphere, extending from about 50 miles upward to a few hundred miles, which is ionized by solar x-rays and ultraviolet lights; it can reflect short radio waves.

magnetic midnight meridian: the extension of the magnetic noon meridian from the geomagnetic pole to the night sector.

magnetic noon meridian: an imaginary line connecting the geomagnetic pole and the Sun.

magnetosphere: the magnetic cavity that surrounds Earth and diverts the flow of solar wind, keeping the wind at a distance of about 40,000 miles from the planet's surface.

nanometer (nm): one-billionth of a meter.

patches: flickering bright spots at the bottom edge of a faint auroral curtain.

photosphere: the visible disk of the sun.

plasma: ionized gas (split into its constituent protons and electrons).

rayed arc: an auroral curtain that develops fine pleats that look brighter than the rest of the structure.

red line: atomic oxygen emission, with a wavelength of 630.0 nm, that occurs above 150 miles in altitude — the "bloody red color" that caused so much fear in medieval days.

red lower border: the bottom folded curtain of an active aurora that is often tinted a crimson red, emitted by excited nitrogen molecules (N_2).

solar flare: an intense transient solar activity.

solar wind: the flow of gases mostly composed of ionized hydrogen atoms (namely, protons and electrons), caused by expansion of part of the Sun's corona; the flow extends to the edges of the solar system.

space weather prediction: the prediction by auroral scientists and solar physicists of phenomena such as solar wind speed variations used to predict auroral activity.

sunspot: a dark spot on the solar disk as viewed from Earth; sunspots are part of the solar cycle, which determines auroral activity.

sunspot cycle: an approximately 11-year span during which sunspots wax and wane.

Van Allen radiation belts: Two zones of high-intensity radiation that remain fixed in Earth's magnetic field and surround the planet at an altitude beginning about 500 miles and reaching tens of thousands of miles into space.

westward traveling surge: a westward moving bulge in the auroral oval during an auroral substorm. ■

Bibliography

Akasofu, Syun-Ichi, Benson Fogle, and Bernhard Haurwitz, eds. *Sydney Chapman, Eighty; from His Friends.* Boulder, Colo.: National Center for Atmospheric Research, 1968.

Amundsen, Roald. *The South Pole: An Account of the Norwegian Antarctic Expedition in the "Fram," 1910-1912.* Translated by A.G. Chater. London: J[ohn] Murray, 1913.

—, and Lincoln Ellsworth. *First Crossing of the Polar Sea.* New York: Doran, 1927.

Angot, Alfred. *The Aurora Borealis.* New York: D. Appleton and Co., 1897.

Camp, Frank B. *Alaska Nuggets.* Anchorage: Alaska Publishing Co., 1922.

Capron, John Rand. *Aurorae: Their Characters and Spectra.* London: Spon, 1879.

Chapman, Sydney. "History of Aurora and Airglow." In *Proceedings of the NATO Advanced Study Institute,* Aurora and Airglow, *University of Keele, England, August 15-26, 1966,* ed. by B.M. McCormac. New York: Reinhold Publishing Corp., 1967.

Cook, Frederick. *My Attainment of the Pole.* New York: The Polar Publishing Co., 1911.

Cook, James. *A Voyage Towards the South Pole and Round the World.* Vol. 1. London: W. Strahan and T. Cadell, 1777.

Davis, Neil. *The Aurora Watcher's Handbook.* Fairbanks: University of Alaska Press, 1992.

Denzel, Justin F. *Adventure North, The Story of Fridtjof Nansen.* London: Abelard-Schuman, 1968.

Devik, Olav. "Kristian Birkeland as I Knew Him." In the *Birkeland Symposium on Aurora and Magnetic Storms.* Sandefjord, Norway: Centre National de la Recherche Scientifigue, 1968.

Ellis, Edward S. *Among the Esquimaux; or Adventures Under the Arctic Circle.* Philadelphia: The Penn Publishing Co., 1894.

Forbes, Charles S. *Iceland: Its Volcanoes, Geysers, and Glaciers.* London: J[ohn] Murray, 1860.

Franklin, Benjamin. *The Writings of Benjamin Franklin.* Ingenhousz, 1778.

Franklin, Sir John. *Narrative of a Journey to the Shores of the Polar Sea, in the Years 1819, 1820, 1821, and 1822.* London: John Murray, 1823.

Gerson, N.C., T.J. Keneshea, and R.J. Donaldson Jr., eds. *Proceedings on the Conference on Auroral Physics.* TR 54-203, Geophysical Research Papers No. 30. Air Force Cambridge Research Center, 1954.

Gray, Louis Herbert, ed. *The Mythology of All Races.* New York: Cooper Square Publishers, 1964.

Greely, Adolphus Washington. *Three Years of Arctic Service: an Account of the Lady Franklin Bay Expedition of 1881-84 and the Attainment of the Farthest North.* New York: C. Scribner's Sons, 1894.

Hall, Charles F. *Arctic Researches and Life Among the Esquimaux: Narrative of an Expedition in Search of Sir John Franklin, in the Years 1860, 1861, and 1862.* London: Sampson, Low, Son and Marston, 1864.

Hawkes, Ernest W. *The Labrador Eskimo.* Ottawa: Government Printing Bureau, 1916.

Huxley, Leonard, arranger. *Scott's Last Expedition, The Personal Journals of Captain R.F. Scott, R.N., C.V.O. on his Journey to the South Pole.* London: John Murray, 1941.

Hooper, Lt. W[illiam] H[ulme]. *Ten Months Among the Tents of the Tuski, with Incidents of an Arctic Boat Expedition in Search of Sir John Franklin*. London: J[ohn] Murray, 1853.

Judson, Katherine Berry, selections by. *Myths and Legends of the Pacific Northwest, Especially of Washington and Oregon*. 4th ed. of facsimilie reprint. Chicago: A.C. McClurg, 1910.

Kane, Elisha Kent. *U.S. Grinnell Expedition in Search of Sir John Franklin*. New York: Harper and Brothers, 1854.

Kennan, George. *Tent Life in Siberia, and Adventures Among the Koraks and Other Tribes in Kamchatka and Northern Asia*. New York: G.P. Putnam and Sons, 1870.

Knox, Thomas W. *The Voyage of the Vivian*. New York: Harper and Brothers, 1884.

Larson, Laurence Marcellus, translator. *The King's Mirror*. Translated from the Old Norwegian. New York: The American-Scandinavian Foundation, 1917.

Lomonosov, Mikhail Vasil'evich. *But, where, O Nature, is thy law? …* Translated by K. Chapman and quoted in Sydney Chapman's *History of Aurora and Airglow*, in *Proceedings of the NATO Advanced Study Institute, Aurora and Airglow, University of Keele, England, August 15-26, 1966*, ed. by B.M. McCormac. New York: Reinhold Publishing Corp., 1967.

London, Jack. *The Call of the Wild*. New York: MacMillan, 1927.

Lynch, Jeremiah. *Three Years in the Klondike*. London: E[dward] Arnold, 1904.

M'Hardie, Elizabeth. *The Midnight Cry: "behold the bridegroom cometh."* London: S.W. Partridge, 1883.

Neatby, Leslie H. *In Quest of the Northwest Passage*. New York: Thomas Y. Crowell Company, 1958.

Nordenskiöld, A[dolf] E. *The Voyage of the Vega Round Asia and Europe*. London: Macmillan and Co., 1881.

Parry, William E. *Journal of a Second Voyage for the Discovery of a North-West Passage from the Atlantic to the Pacific; Performed in the Years 1821-22-23, in His Majesty's Ships Fury and Hecla*. N.p., 1904.

Peary, Robert E. *The North Pole* London: Hodder and Stoughton, 1910.

Ramsay, William. *The Gases of the Atmosphere: The History of their Discovery*. 3rd ed. London: Macmillan and Co., 1905.

Rasmussen, Knud. *Intellectual Culture of the Iglulik Eskimo*. Fifth Thule Expedition. Copenhagen: Glydem Dalski Boghand, 1932.

Royal Society. *Biographical Memoirs of Fellows of the Royal Society*. Vol. 1. London: Royal Society, 1955.

Scearce, Stanley. *Northern Lights to Fields of Gold*. Illustrated by R.H. Hall. Caldwell, Idaho: Caxton Printers, 1939.

Service, Robert. *Collected Poems of Robert Service*. New York: Dodd, Mead, 1940.

Silverman, Sam M., and Tai-Fu Tuan. *Auroral Audibility*. Edited by Helmut Erich Landsberg and J. Van Meighen. Advance in Geophysics, Vol. 16. New York: Academic Press, 1973.

Störmer, Carl. *The Polar Aurora*. Oxford: Clarendon Press, 1955.

Taylor, Bayard. *Prose Writings of Bayard Taylor. …* New York: G.P. Putnam, 1862.

Tromholt, Sophus. *Under the Rays of the Aurora Borealis*. Boston: Houghton, Mifflin, 1885.

Whymper, Frederick. *Travel and Adventure in the Territory of Alaska*. London: J[ohn] Murray, 1868.

WEB SITES

http://dac3.pfrr.alaska.edu/aurora/ (Poker Flat Research Range Aurora Information and Images)

www.alaskascience.com (Alaska Science Explained)

www.exploratorium.edu/learning_studio/auroras/ (Auroras: Painting in the Sky)

www.geo.mtu.edu/weather/aurora/ (The Aurora Page)

www.gi.alaska.edu/cgi-bin/predict.cgi (Aurora Forecast from the Geophysical Institute at the University of Alaska Fairbanks)

www.sec.noaa.gov/pmap (POES [satellite] Auroral Activity)

www.sec.noaa.gov/today.html (Today's Space Weather)

www.spaceweather.com (Spaceweather at NASA)

Index

CONTRIBUTORS

Membership in The Alaska Geographic Society includes a subscription to *ALASKA GEOGRAPHIC*®, the Society's colorful, award-winning quarterly. Contact us for current membership rates or to request a free catalog.

The *ALASKA GEOGRAPHIC*® back issues listed below can be ordered directly from us. NOTE: This list was current in early 2002. If more than a year has elapsed since that time, be sure to contact us before ordering to check prices and availability of back issues, particularly for books marked "Limited."

When ordering back issues please add shipping: $5 for the first book and $2 for each additional book. Inquire for shipping rates to non-U.S. addresses. To order, send check or money order (U.S. funds) or VISA or MasterCard information (including expiration date and daytime phone number) with list of titles desired to:

ALASKA GEOGRAPHIC.

P.O. Box 93370 • Anchorage, AK 99509-3370
Phone (907) 562-0164 • Toll free (888) 255-6697
Fax (907) 562-0479 • e-mail: info@akgeo.com

The North Slope, Vol. 1, No. 1. Out of print.
One Man's Wilderness, Vol. 1, No. 2. Out of print.
Admiralty...Island in Contention, Vol. 1, No. 3. $9.95.
Fisheries of the North Pacific, Vol. 1, No. 4. Out of print.
Alaska-Yukon Wild Flowers, Vol. 2, No. 1. Out of print.
Richard Harrington's Yukon, Vol. 2, No. 2. Out of print.
Prince William Sound, Vol. 2, No. 3. Out of print.
Yakutat: The Turbulent Crescent, Vol. 2, No. 4. Out of print.
Glacier Bay: Old Ice, New Land, Vol. 3, No. 1. Out of print.
The Land: Eye of the Storm, Vol. 3, No. 2. Out of print.
Richard Harrington's Antarctic, Vol. 3, No. 3. $9.95.
The Silver Years, Vol. 3, No. 4. $21.95. Limited.
Alaska's Volcanoes, Vol. 4, No. 1. Out of print.
The Brooks Range, Vol. 4, No. 2. Out of print.
Kodiak: Island of Change, Vol. 4, No. 3. Out of print.
Wilderness Proposals, Vol. 4, No. 4. Out of print.
Cook Inlet Country, Vol. 5, No. 1. Out of print.
Southeast: Alaska's Panhandle, Vol. 5, No. 2. Out of print.
Bristol Bay Basin, Vol. 5, No. 3. Out of print.
Alaska Whales and Whaling, Vol. 5, No. 4. $19.95.
Yukon-Kuskokwim Delta, Vol. 6, No. 1. Out of print.
Aurora Borealis, Vol. 6, No. 2. $21.95. Limited
Alaska's Native People, Vol. 6, No. 3. $29.95. Limited.
The Stikine River, Vol. 6, No. 4. $9.95.
Alaska's Great Interior, Vol. 7, No. 1. $19.95.
Photographic Geography of Alaska, Vol. 7, No. 2. Out of print.
The Aleutians, Vol. 7, No. 3. Out of print.
Klondike Lost, Vol. 7, No. 4. Out of print.
Wrangell-Saint Elias, Vol. 8, No. 1. Out of print.

Alaska Mammals, Vol. 8, No. 2. Out of print.
The Kotzebue Basin, Vol. 8, No. 3. Out of print.
Alaska National Interest Lands, Vol. 8, No. 4. $19.95.
*Alaska's Glaciers, Vol. 9, No. 1. Rev. 1993. $21.95. Limited.
Sitka and Its Ocean/Island World, Vol. 9, No. 2. Out of print.
Islands of the Seals: The Pribilofs, Vol. 9, No. 3. $9.95.
Alaska's Oil/Gas & Minerals Industry, Vol. 9, No. 4. $9.95.
Adventure Roads North, Vol. 10, No. 1. $9.95.
Anchorage and the Cook Inlet Basin, Vol. 10, No. 2. $19.95.
Alaska's Salmon Fisheries, Vol. 10, No. 3. $9.95.
Up the Koyukuk, Vol. 10, No. 4. $9.95.
Nome, Vol. 11, No. 1. $21.95. Out of print.
Alaska's Farms and Gardens, Vol. 11, No. 2. $19.95.
Chilkat River Valley, Vol. 11, No. 3. $9.95.
Alaska Steam, Vol. 11, No. 4. $19.95.
Northwest Territories, Vol. 12, No. 1. $9.95.
Alaska's Forest Resources, Vol. 12, No. 2. $9.95.
Alaska Native Arts and Crafts, Vol. 12, No. 3. $24.95.
Our Arctic Year, Vol. 12, No. 4. $19.95.
* Where Mountains Meet the Sea, Vol. 13, No. 1. $19.95.
Backcountry Alaska, Vol. 13, No. 2. $9.95.
British Columbia's Coast, Vol. 13, No. 3. $9.95.
Lake Clark/Lake Iliamna, Vol. 13, No. 4. Out of print.
Dogs of the North, Vol. 14, No. 1. Out of print.
South/Southeast Alaska, Vol. 14, No. 2. $21.95. Limited.
Alaska's Seward Peninsula, Vol. 14, No. 3. $19.95.
The Upper Yukon Basin, Vol. 14, No. 4. $19.95.
Glacier Bay: Icy Wilderness, Vol. 15, No. 1. Out of print.
Dawson City, Vol. 15, No. 2. $19.95.
Denali, Vol. 15, No. 3. $9.95.
The Kuskokwim River, Vol. 15, No. 4. $19.95.
Katmai Country, Vol. 16, No. 1. $19.95.
North Slope Now, Vol. 16, No. 2. $9.95.
The Tanana Basin, Vol. 16, No. 3. $9.95.
* The Copper Trail, Vol. 16, No. 4. $19.95.
* The Nushagak Basin, Vol. 17, No. 1. $19.95.
* Juneau, Vol. 17, No. 2. Out of print.
* The Middle Yukon River, Vol. 17, No. 3. $19.95.
* The Lower Yukon River, Vol. 17, No. 4. $19.95.
* Alaska's Weather, Vol. 18, No. 1. $9.95.
* Alaska's Volcanoes, Vol. 18, No. 2. $19.95. Limited
Admiralty Island: Fortress of Bears, Vol. 18, No. 3. Out of print.
Unalaska/Dutch Harbor, Vol. 18, No. 4. Out of print.
* Skagway: A Legacy of Gold, Vol. 19, No. 1. $9.95.
Alaska: The Great Land, Vol. 19, No. 2. $9.95.
Kodiak, Vol. 19, No. 3. Out of print.
Alaska's Railroads, Vol. 19, No. 4. $19.95.
Prince William Sound, Vol. 20, No. 1. $9.95.
Southeast Alaska, Vol. 20, No. 2. $19.95.
Arctic National Wildlife Refuge, Vol. 20, No. 3. $19.95.
Alaska's Bears, Vol. 20, No. 4. $19.95.
The Alaska Peninsula, Vol. 21, No. 1. $19.95.

The Kenai Peninsula, Vol. 21, No. 2. $19.95.
People of Alaska, Vol. 21, No. 3. $19.95.
Prehistoric Alaska, Vol. 21, No. 4. $19.95.
Fairbanks, Vol. 22, No. 1. $19.95.
The Aleutian Islands, Vol. 22, No. 2. $19.95.
Rich Earth: Alaska's Mineral Industry, Vol. 22, No. 3. $19.95.
World War II in Alaska, Vol. 22, No. 4. $19.95.
Anchorage, Vol. 23, No. 1. $21.95.
Native Cultures in Alaska, Vol. 23, No. 2. $19.95.
The Brooks Range, Vol. 23, No. 3. $19.95.
Moose, Caribou and Muskox, Vol. 23, No. 4. $19.95.
Alaska's Southern Panhandle, Vol. 24, No. 1. $19.95.
The Golden Gamble, Vol. 24, No. 2. $19.95.
Commercial Fishing in Alaska, Vol. 24, No. 3. $19.95.
Alaska's Magnificent Eagles, Vol. 24, No. 4. $19.95.
Steve McCutcheon's Alaska, Vol. 25, No. 1. $21.95.
Yukon Territory, Vol. 25, No. 2. $21.95.
Climbing Alaska, Vol. 25, No. 3. $21.95.
Frontier Flight, Vol. 25, No. 4. $21.95.
Restoring Alaska: Legacy of an Oil Spill, Vol. 26, No. 1. $21.95.
World Heritage Wilderness, Vol. 26, No. 2. $21.95.
The Bering Sea, Vol. 26, No. 3. $21.95.
Russian America, Vol. 26, No. 4, $21.95
Best of *ALASKA GEOGRAPHIC*®, Vol. 27, No. 1, $24.95
Seals, Sea Lions and Sea Otters, Vol. 27, No. 2, $21.95
Painting Alaska, Vol. 27, No. 3, $21.95
Living Off the Land, Vol. 27, No. 4, $21.95
Exploring Alaska's Birds, Vol. 28, No. 1, $23.95
Glaciers of Alaska, Vol. 28, No. 2, $23.95
Inupiaq and Yupik People of Alaska, Vol. 28, No. 3, $23.95
The Iditarod, Vol. 28, No. 4, $23.95

* Available in hardback (library binding) — $24.95 each.

PRICES AND AVAILABILITY SUBJECT TO CHANGE

NEXT ISSUE	Vol. 29, No. 2

Boating Alaska

This book illustrates one of the best ways to see Alaska — by non-motorized watercraft. Modern-day canoers, kayakers, rafters, and sailors present unforgettable journeys down the state's great rivers and along its thousands of miles of fresh- and saltwater coastline. Visit waters of the Inside Passage, Kenai Fjords National Park, the Wood-Tikchik system, the Aleutians, the mighty Yukon River, and more. To members summer 2002.